本书由中国科学院数学与系统科学研究院资助出版

数学 24/7

体育中的数学

〔美〕雷·西蒙斯　著

刘　刚　译

科学出版社

北　京

图字：01-2015-5621号

内 容 简 介

体育中的数学是"数学生活"系列之一，内容涉及众多游戏的计分和规则，如棒球的得分、击球率和防御率，篮球的得分及球员身高统计，橄榄球的得分及罗马数字识别，足球的禁区，网球的得分，以及奥运奖牌统计与分析等，让青少年在学校学到的数学知识应用到与体育有关的多个方面，让青少年进一步了解数学在日常生活中是如何运用的。

本书适合作为中小学生的课外辅导书，也可作为中小学生的兴趣读物。

Copyright © 2014 by Mason Crest, an imprint of National Highlights, Inc. All rights reserved. No part of this publication may be reproduced or transmitted in any form or by any means, electronic or mechanical, including photocopying, recording, taping or any information storage and retrieval system, without permission from the publisher.

The simplified Chinese translation rights arranged through Rightol Media.
（本书中文简体版权经由锐拓传媒取得Email:copyright@rightol.com）

图书在版编目（CIP）数据

体育中的数学/（美）雷·西蒙斯（Rae Simons）著；刘刚译.—北京:科学出版社,2018.5
（数学生活）
书名原文：Sport Math
ISBN 978-7-03-056747-5

Ⅰ.①体… Ⅱ.①雷…②刘… Ⅲ.①数学-青少年读物 Ⅳ.①O1-49

中国版本图书馆CIP数据核字（2018）第046673号

责任编辑:胡庆家 / 责任校对:邹慧卿
责任印制:肖 兴 / 封面设计:陈 敬

科学出版社 出版
北京东黄城根北街16号
邮政编码：100717
http://www.sciencep.com

北京汇瑞嘉合文化发展有限公司 印刷
科学出版社发行 各地新华书店经销
*
2018年5月第 一 版　　开本：884×1194 1/16
2018年5月第一次印刷　　印张:4 1/2
字数：70 000

定价：98.00元（含2册）
（如有印装质量问题，我社负责调换）

引　　言

你会如何定义数学？它也许不是你想象的那样简单。我们都知道数学和数字有关。我们常常认为它是科学，尤其是自然科学、工程和医药学的一部分，甚至是基础部分。谈及数学，大多数人会想到方程和黑板、公式和课本。

但其实数学远不止这些。例如，在公元前5世纪，古希腊雕刻家波留克列特斯曾经用数学雕刻出了"完美"的人体像。又例如，还记得列昂纳多·达·芬奇吗？他曾使用有着赏心悦目的尺寸的几何矩形——他称之为"黄金矩形"，创作出了著名的画作——蒙娜丽莎。

数学和艺术？是的！数学对包括医药和美术在内的诸多学科都至关重要。计数、计算、测量、对图形和物理运动的研究，这些都被融入到音乐与游戏、科学与建筑之中。事实上，作为一种描述我们周围世界的方式，数学形成于日常生活的需要。数学给我们提供了一种去理解真实世界的方法——继而用切实可行的途径来控制世界。

例如，当两个人合作建造一样东西时，他们肯定需要一种语言来讨论将要使用的材料和要建造的对象。但如果他们建造的过程中没有用到一个标尺，也不用任何方式告诉对方尺寸，甚至他们不能互相交流，那他们建造出来的东西会是什么样的呢？

事实上，即便没有察觉到，但我们确实每天都在使用数学。当我们购物、运动、查看时间、外出旅行、出差办事，甚至烹饪时都用到了数学。无论有没有意识到，我们在数不清的日常活动中用着数学。数学几乎每时每刻都在发生。

很多人都觉得自己讨厌数学。在我们的想象中，数学就是枯燥乏味的老教授做着无穷无尽的计算。我们会认为数学和实际生活没有关系；离开了数学课堂，在真实世界里我们再不用考虑与数学有关的事情了。

然而事实却是数学使我们生活各方面变得更好。不懂得基本的数学应用的人会遇到很多问题。例如，美联储发现，那些破产的人的负债是他们所得收入的1.5倍左右——换句话说，假设他们年收入是24000美元，那么平均负债是36000美元。懂得基本的减法，会使他们提前意识到风险从而避免破产。

作为一个成年人，无论你的职业是什么，都会或多或少地依赖于你的数学计算能力。没有数学技巧，你就无法成为科学家、护士、工程师或者计算机专家，就无法得到商学院学位，就无法成为一名服务生、一位建造师或收银员。

体育运动也需要数学。从得分到战术，都需要你理解数学——所以无论你是想在电视上看一场足球比赛，还是想在赛场上成为一流的运动员，数学技巧都会给你带来更好的体验。

还有计算机的使用。从农庄到工厂、从餐馆到理发店，如今所有的商家都至少拥有一台电脑。千兆字节、数据、电子表格、程序设计，这些都要求你对数学有一定的理解能力。当然，电脑会提供很多自动运算的数学函数，但你还得知道如何使用这些函数，你得理解电脑运行结果的含义。

这类数学是一种技能，但我们总是在需要做快速计算时才会意识到自己需要这种技能。于是，有时我们会抓耳挠腮，不知道如何将学校里学的数学应用在实际生活中。这套丛书将助你一马当先，让你提前练习数学在各种生活情境里的运用。这套丛书将会带你入门——但如果想掌握更多，你必须专心上数学课，认真完成作业，除此之外再无捷径。

但是，付出的这些努力会在之后的生活里——几乎每时每刻（24/7）——让你受益匪浅！

目　　　录

引言
1. 棒球：打击率　　　　　　　　　　　　　　　　1
2. 棒球：投手防御率　　　　　　　　　　　　　　3
3. 棒球：得分　　　　　　　　　　　　　　　　　5
4. 篮球：平均身高　　　　　　　　　　　　　　　7
5. 篮球：胜-负　　　　　　　　　　　　　　　　　9
6. 橄榄球：得分　　　　　　　　　　　　　　　　11
7. 橄榄球：罗马数字　　　　　　　　　　　　　　13
8. 足球：禁区　　　　　　　　　　　　　　　　　15
9. 网球：得分　　　　　　　　　　　　　　　　　17
10. 径赛：纪录　　　　　　　　　　　　　　　　　19
11. 高尔夫球：标准杆　　　　　　　　　　　　　　21
12. 游泳：圈数和距离　　　　　　　　　　　　　　23
13. 曲棍球：角度　　　　　　　　　　　　　　　　25
14. 奥运会奖牌　　　　　　　　　　　　　　　　　27
15. 小结　　　　　　　　　　　　　　　　　　　　29
参考答案　　　　　　　　　　　　　　　　　　　　32

Contents

INTRODUCTION
1. BASEBALL: BATTING AVERAGES 37
2. BASEBALL: EARNED RUN AVERAGES 39
3. BASEBALL: SCORING 40
4. BASKETBALL: AVERAGE HEIGHT 42
5. BASKETBALL: WIN-LOSS 43
6. FOOTBALL: SCORING 45
7. FOOTBALL: ROMAN NUMERALS 46
8. SOCCER: FIELD AREA 47
9. TENNIS: SCORING 49
10. TRACK: RECORDS 50
11. GOLF: PAR 52
12. SWIMMING: LAPS AND DISTANCE 53
13. HOCKEY: ANGLES 55
14. OLYMPIC MEDALS 56
15. PUTTING IT ALL TOGETHER 58
ANSWERS 59

1
棒球：打击率

杰伊喜欢运动，他在秋季、冬季和春季参加了学校的多项运动。与此同时，杰伊在电视、网络和报纸上也关注着一切与运动有关的信息。在众多喜欢的运动中，杰伊最关注棒球。杰伊也喜欢数学。棒球运动中涉及很多数学知识。球迷和专业人士通过数据统计分析来评价球员的表现，并用数学工具来预测在比赛中可能发生的事情。

棒球中常使用的一个数字指标是打击率。打击率是衡量一个击球手击出安打的频率。击球手一直击球不中出局将不会有好的打击率。反之，击球手常能击球进垒(单、双、三或全垒打)将有很好的打击率。只有击球中垒才能算分。下面我们将介绍如何计算打击率。

打击率如0.320, 0.245, 0.222，是小数形式下的百分比值。为了理解打击率的百分比形式，我们简单向右移动两位小数点。这个数字就是选手上场击球获得垒打的百分比值。也就是说打击率为0.320表示击中球的概率为32%，0.245的打击率有24.5%击球成功率。

1. 以下打击率的百分比形式是什么？

    0.198 =
    0.304 =
    0.271 =
    0.336 =

打击率的计算公式：

$$打击率 = 击中次数 \div 击球次数$$

如果选手在场上击球30次，击中垒9次，那么他的打击率为

$$9次 \div 30次 = 0.300$$

击球率总是保留三个数字，所以必须将打击率结果计算到千分位(小数点右边的第三位数字)。

2. 如果杰伊最喜欢的球员在上赛季击球540次，中垒175次，那么他的打击率为多少？

如果你知道了他的打击率和在场上击球次数，也可以计算出中垒次数。只需要简单地重新排列方程：

$$击中次数 = 打击率 \times 击球次数$$

3. 如果选手的打击率为0.322，场上击球193次，那么中垒次数为多少？

现代棒球运动中，选手的打击率几乎没有高于0.400的。但是打击率低于0.200时被认为是低水平的。

4. 如果选手场上353次击球中有77次中垒，那么他的水平是否在这个区间内？他的打击率是多少？

2
棒球：投手防御率

在观看棒球直播时，杰伊特别关注投手。因为在小的时候，他曾经是少年棒球联合会（美）中的一个投手。在和朋友们玩棒球时，他也很喜欢做投手。

击球率告诉杰伊一个击球手的水平有多高，投手防御率告诉他一个投手的水平有多高。投手防御率是衡量投手所失的自责分。防御率越低投手水平越高，因为这意味着投手不让对手得分。下面我们就介绍如何计算投手防御率。

为计算一个投手在一场比赛中的防御率，首先合计他投出的局数。投手一般不投满整个比赛，所以他可能只投2，5或8局。另外还得计算部分局。比如投手投了1/3局，那么加上0.33；投了2/3局，那么加上0.67。

现在，对手在这些局中的累计得分，就是投手的自责分。自责分乘以9。

最后，这个数字除以投手投的局数。确保小数点后只有两个数字。整个方程是这样的：

投手防御率 =（自责分 × 9）÷ 局数

1. 投手在比赛中投了5局，自责分是2分，那么投手的防御率是多少？

用同样的方法可以计算投手在整个赛季、职业生涯中的防御率。区别仅仅是使用更大的数字而已。

2. 上赛季，某投手投了325局，自责分是89分，那么他的防御率是多少？

就像击球率一样，可以通过变换防御率计算公式来求解其他项。比如，如果知道投手的防御率和投球局数，就可以计算出他的自责分。

3. 投手上月投了56.67局，防御率为4.2，那么他上个月自责分是多少？自责分应该是一个整数，

自责分 =（防御率 × 局数）÷ 9

因此

自责分 =（4.2 × 56.67）÷ 9
自责分 =

防御率低于2.00是令人惊讶的水平，但高于5.00表明其水平比较低，面临被淘汰。

4. 这里有一些防御率数值：3.78, 4.89, 1.93, 2.55, 5.09。对防御率按投手水平从低到高排序。

3
棒球：得分

杰伊观看棒球比赛时喜欢填写计分卡，他想了解队伍是如何得分的，尽管在记分牌上也能看到得分。

棒球比赛中用到很多数字。1次出局有3振，1个四坏保送有4个坏球。1个回合有3次出局，一场比赛有9个回合。另外还有一垒安打、二垒安打、三垒安打和全垒打。一垒安打是指球手率先跑到一垒。如果任何其他队员已上垒，那么他们也要先到达一垒。二垒安打要求队员率先跑到二垒，同时先于其他队员一垒或者二垒，这基本取决于他们能到达多远。击中三垒要求队员率先跑到第三个垒且先于其他队员至少一垒。击中全垒打要求队员跑回本垒且其他队员上垒！

杰伊发现填写计分卡使得跟踪了解这些数据更容易！下面，我们看看如何记录一场比赛。

下面是杰伊最喜欢的球队近期某场比赛中发生的事情。

第一回合：1号—3振；2号——垒安打；3号—3振；4号—二垒安打（2号领先二垒）；5号——垒安打（2号首先回到本垒，4号在二垒处出）

第二回合：1号—飞球；2号—4个坏球；3号—3振；4号—3振

第三回合：1号—3振；2号—飞球；3号—全垒打；4号—飞球

第四回合：1号—4个坏球；2号——垒安打（1号领先1垒）；3号——垒安打（1号领先1垒，2号领先1垒）；4号—二垒安打（1号先到本垒；2号在3垒出局；3号在2垒出局）；5号—3振

第五回合：1号——垒安打；2号—飞球；3号—3振；4号—3振

第六回合：1号—3振；2号—二垒安打；3号—3振；4号——垒安打（2号领先到达3垒）；5号—飞球

第七回合：1号—3振；2号—3振；3号—3振

第八回合：1号—飞球；2号—飞球；3号—二垒安打；4号—3振

第九回合：1号—3振；2号—本垒打；3号——垒安打；4号—飞球

1. 有多少次3振？

2. 有多少个四坏保送？

3. 球员多少次安全到达三垒或更远？

4. 最后的分数是多少？如果对手有3分，杰伊的球队赢了还是输了？

4
篮球：平均身高

杰伊也喜欢篮球。他喜欢在电视上看比赛或者玩游戏。只要有时间，他就和朋友们一起打篮球。

杰伊从未在学校里面打篮球，主要原因是他不够高。篮球运动员总是比较高，这样可以更容易地够着篮筐。无论高矮胖瘦都可以享受篮球乐趣甚至打得很好，但教练通常要高的球员。

大学和职业篮球运动员——不论是男的还是女的——一般都比较高。下面我们将计算不同的平均高度，包括平均值、中位数和众数。

以下是两届美国女子篮球联盟冠军：西雅图风暴队2012年在册队员的身高情况：

6' 2" (74英寸)
5' 9" (69英寸)
5' 10" (70英寸)
6' 6" (78英寸)
6' 3" (75英寸)
6' 2" (74英寸)
5' 11" (71英寸)
6' 2" (74英寸)
6' 2" (74英寸)
6' 4" (76英寸)
5' 11" (71英寸)

通常说的平均指的是平均值。需要将所有人的身高值相加，然后除以人数。只用英寸计算平均数比较容易，然后再转化为几英尺几英寸。

1. 他们身高平均多少英寸？精确到英寸。换算成几英尺几英寸是多少？

中位数是另外一种平均值。首先将身高值从小到大排序，然后删除最小值和最大值，一直删下去直到只剩下最后一个数字。如果剩下两个数字，计算平均值就找到了精确的中位数。

2. 这些身高值的中位数是多少？

第三种平均是众数。众数是身高数值中出现次数最多的数值。

3. 众数是多少？

4. 如果美国女性的平均身高约5英尺4英寸，那么西雅图风暴队女球员的平均身高比5英尺4英寸高多少英寸？

5
篮球：胜-负

篮球赛季期间，杰伊喜欢一直关注着高校联赛和职业联赛的每支球队。他每周在网上查看球队的排名，看胜负数。当他看到9-3，就知道球队赢了9场，输了3场，一共打了12场比赛。

另一个与胜负数有关的数字是胜率。胜率是球队获胜场数与比赛总场数的比值。看看每个球队的胜率，就可以知道这个赛季目前哪支球队发挥最好。下面我们将了解胜负数和胜率。

杰伊看了赛季期间几支球队的胜负数：

4-10
14-1
5-10
9-5
7-7
3-12
1-13

1. 球队胜负数1-13和14-1相比，那支球队的胜负记录好一些？

2. 这两支球队打了多少场比赛？

有时候两支球队比赛的场次不同，那么比较就需要一些技巧。为了比较这两支球队，需要计算其胜率，公式：

$$胜率=赢球场数/比赛场数$$

在篮球比赛中，胜率通常表示成小数，保留三位小数。要想转换为百分数，只需将小数点向右移动两位。

3. 球队比分1-13和14-1的胜率分别是多少？给出小数和百分比形式的答案。

4. 将球队按照胜率大小从高到低排序：

    A.
    B.
    C.
    D.
    E.
    F.
    G.

6
橄榄球：得分

另一项杰伊喜欢的运动是橄榄球。他常在电视上看球赛，周末和朋友一起玩，甚至参加学校的比赛。他梦想将来成为一名职业橄榄球运动员。

橄榄球和其他运动一样，涉及很多数字和数学知识。任何观看比赛或者踢球的人都需要学会各种计分方式。每场比赛的情况都不一样，即使最终比分是一样的。下面我们将计算如何累加各种得分得到总分。

这是橄榄球比赛中球员得分情况：

触地=6分

触地附加=1分

2分转换=2分

射门得分=3分

安全得分=2分

在杰伊最后参加的橄榄球比赛中，他的高中队赢了34分，对方球队得了29分。

1. 杰伊所在球队有4次触地得分。那么除了触地得分之外他们得了多少分？

2. 如果其他4种方式都得分，给出得到这个分数的可能情况？

3. 如果负队也有4次触地得分，那么他们得到29分的一种可能情况是什么？

4. 如果对手多得到一次触地得分，他们会赢得比赛吗？为什么？

7

橄榄球：罗马数字

超级碗比赛日是杰伊最喜欢的一天。他和家人一起观看比赛，制作特别的食物，穿球队的衣服。

2013年，他们观看了第47届超级碗比赛。然而比赛公告使用的不是我们在数学课上使用的数字。超级碗的届数使用罗马数字表达，因此2013届写成超级碗XLVII。杰伊的妹妹困惑于如何阅读罗马数字，因此杰伊在比赛中常教她。我们在下面具体学习：

在罗马数字中，不同字母代表特定的数字：

$$I = 1$$
$$V = 5$$
$$X = 10$$
$$L = 50$$
$$C = 100$$
$$D = 500$$
$$M = 1000$$

通过排列不同的字母构成数字。2就是两个I合在一起：II。3就是三个I合在一起：III。但4有点不同，不是连续四个I排成一列，而是IV，表示比5小1。十的表示类似，只是用X代替V。

1. 从1到10用罗马数字怎么写？

第47届超级碗，罗马数字表示为XLVII。你可以先将这个数字分开然后再加在一起。先按位置划分：首先看十位，然后看个位。

十位是XL。意思是比50少10。你有可能将XL想成60，因为10加50正好是60。但是每次看到前面的字母小于后面的字母时，你得使用减法而不是加法。如果这个字母表示60，那么它应该是LX，10出现在50的后面。

XL后面是VII，那是5加上2，也就是7。十位的字母40加上个位的字母7，也就是47。

2. 第46届超级碗用罗马数字怎么表示？

3. 在2023年时，超级碗是第57届。那么57用罗马数字怎么表示？

你也可以给出更大数字的罗马数字形式，比如"第47届超级碗—2013"中的2013。

4. 2013用罗马数字怎么表示？

8
足球：禁区

杰伊的朋友阿那亚喜欢踢足球。她是校队的明星球员之一，希望有一天成为一名职业足球运动员。

阿那亚和杰伊所在的学校正在建造一个新的足球场，因为旧足球场被洪水淹了，而且也太旧了。新的足球场必须精确测量，尺寸很重要。每个学校和职业比赛中的足球场大小都是相同的，球员们也习惯了标准球场的大小。这样不同球队在这个球场上比赛不存在不公平的问题。

下面将带你了解足球场的尺寸，并将帮助你计算场地面积。

学校的新足球场有100码长、60码宽（这里，1码等于3英尺）。你已经有足够的信息来计算球场的面积。矩形的面积公式为

$$面积 = 长 \times 宽$$

1. 场地面积是多少平方码？

用平方英尺计算面积。首先，通过每个尺寸乘以3，将单位从码转换成英尺，然后将那些数字代入面积方程。

2. 场地面积是多少平方英尺？

学校还将在场地上画很多线条，并设置球门。例如，中场线将球场在长度方向上一分为二，并设定客队的一侧。

3. 中场线从场地长度方向上看有多远（英尺）？

半场面积是多少平方英尺？

场地的两端球门前是球门区（小禁区）。球员踢球必须通过这个区域才能进球。

球门区的面积为120平方码。如果球门区的一侧是20码长，你可以通过变换面积公式获得其他未知项（宽度）。新公式是

$$宽 = 面积 \div 长$$

4. 所以，宽度 = 120平方码 ÷ 20码 = ？

5. 球门区的宽度是多少英尺？

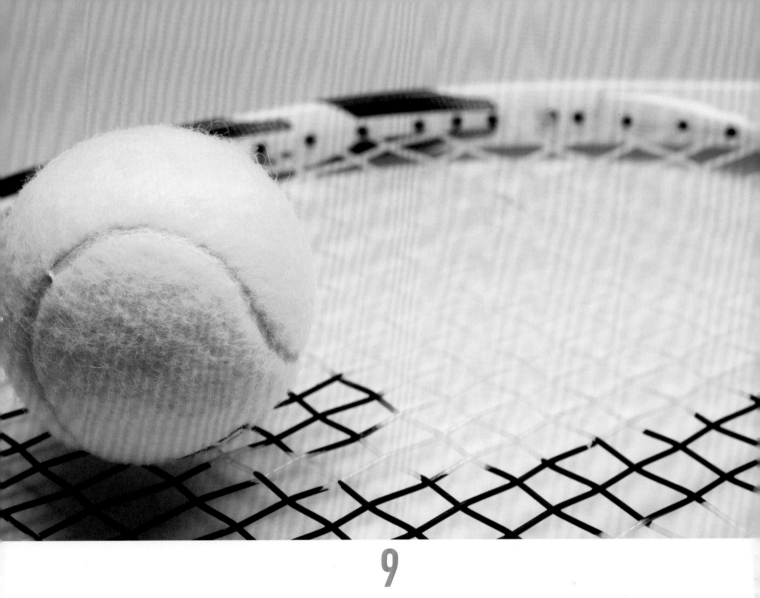

9
网球：得分

杰伊的朋友关打网球。有时，杰伊去看他的比赛并为他呐喊助威。关水平不错——他赢了很多比赛。

起初，杰伊对网球比赛如何计分有点困惑，因为网球不只是从1开始计算点数的。网球使用完全不同的评分系统，不是1,2,3,4这样计分。杰伊很快就明白了，你也可以。

以下是网球计分方法：

0分=LOVE
1分= 15点
2分=30点
3分=40点
4分=60点/比赛结束

这里有一些比赛的计分规则：

- 必须赢得至少4分才能获胜一局。
- 必须多得2分才算获胜一局。
- 如果获得40点的时候没有领先2分，那么再得1分被称为"占优"。如果再获1分，就获胜一局。
- 发球方的分数被首先宣布。

杰伊看过关的某次比赛：关得了1分，对手得了3分。下一局轮到关发球。

1. 此时比赛的比分是多少？

关的对手又获1分。

2. 比分是多少？这一局结束了吗？

然而关和杰伊并不太担心。失去这局并不意味关已经失去了整个比赛。网球比赛中，率先赢得6局比赛并领先对手至少2局才算赢得该盘。随着比赛的继续，关赢了5局比赛。他的对手赢了4局。

3. 如果关赢得下一局比赛，这一盘将会发生什么？

4. 如果关的对手赢得下一局比赛，这盘比赛会发生什么？

最后，网球运动员通常在整个比赛中打3盘。获胜者必须赢得3盘中的2盘。
关和他的对手各赢一盘。他们必须再赛一盘，以决定谁将赢得比赛。随着比赛的继续，最后一盘得分5-4。比赛对关有利。

5. 如果关赢了这盘，关会赢得比赛吗？为什么？

10
径赛：纪录

有一次，杰伊去学校观看田径运动会。他有几个朋友参加了，他喜欢为他们加油。

他想知道径赛的世界纪录。他的朋友速度很快，但他知道，参加训练跑步的男生和女生可以更快！田径运动会结束后，他查找了径赛的世界纪录。他发现他们真的很快，但他不知道如何准确地比较世界纪录和朋友的纪录。

将径赛时间数据放在线图上可以帮助杰伊比较。你可以在下面做同样的事情。

杰伊决定比较最短的径赛——男子100米短跑的纪录。以下是他找到的最近几年世界纪录信息：

2009年世界纪录：9.58秒
2008年世界纪录：9.69秒
2007年世界纪录：9.74秒
2006年世界纪录：9.762秒
2005年世界纪录：9.768秒
2002年世界纪录：9.78秒
1999年世界纪录：9.79秒

这里是他们学校在同年的数据：

2009年学校纪录：10.39秒
2008年学校纪录：10.41秒
2007年学校纪录：10.42秒
2006年学校纪录：10.56秒
2005年学校纪录：10.59秒
2002年学校纪录：10.60秒
1999年学校纪录：10.67秒

现在分两行填写这个线图的其余部分：一个世界纪录，一个学校纪录。

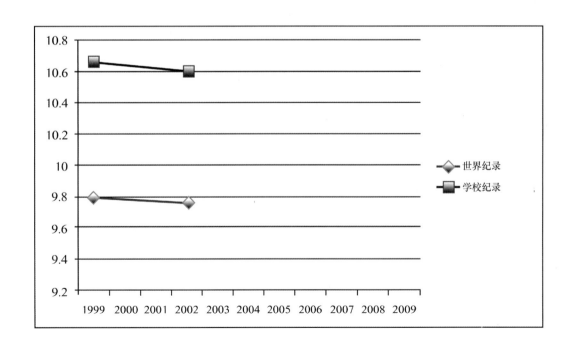

11
高尔夫球：标准杆

杰伊在随意浏览电视频道时看到了高尔夫球比赛。他以前没有真正看过高尔夫球比赛，但这次他决定去了解下如何打高尔夫球，以及如何计分。

他发现，像其他运动一样，高尔夫球也涉及很多数字。在高尔夫球比赛中，球员要经过18个不同的洞。每个洞有一个杆数，即将高尔夫球推入洞内的击球数。较难的洞有较高的杆数，与之相反，较易的洞有较低的杆数。选手通过尽可能少的击球次数击球入洞以获胜。

杰伊很快就知道这些规则，并且可以根据每个洞的分数来判断谁是最好的球员。下面看看你是否能做同样的事情。

杆数是建立在洞的难度上的。高杆数的洞通常距离很远，而且路上有急弯或小山。

1. 一个长200码、直线路径的洞与一个长320码、曲线路径的洞相比，杆数高还是低？为什么？

高尔夫球场通常有18个洞，杆数总计72，但不总都是这样。

2. 一个有4个3杆洞、10个4杆洞和4个5杆洞的球场杆数总计是否为72？

3. 这个高尔夫球场上最困难的洞是哪些？

除了标准杆，针对需要多少次击球才入洞，高尔夫球有很多不同的术语。

双鹰：低于标准杆3杆，或-3
老鹰：低于标准杆2杆，或-2
小鸟：低于标准杆1杆，或-1
标准杆：标准杆或0
柏忌：高于标准杆1杆，或+1
双柏忌：高于标准杆2杆，或+2
三柏忌：高于标准杆3杆，或+3

最好的球员得到小鸟、老鹰或双鹰。

要计算选手的分数，只需将击球入洞次数高于或低于对应洞的标准杆的数字相加即可。

杰伊正在观看的选手在前3个洞中获得一个标准杆、一个柏忌和一个老鹰。

4. 她的总得分是多少？

下一个洞是一个5杆洞，她4次击球入洞了。

5. 她在这个洞上的得分是多少？目前她的总分是多少？

12
游泳：圈数和距离

杰伊的弟弟利奥参加了游泳课，他的最终目标是进校游泳队，但是他首先需要更多的训练。现在，他在一个25码的游泳池里参加培训。如果他进入了游泳队，他将在一个更大的、50米的游泳池游泳。

利奥在池中游的每一圈都是固定的距离。游泳比赛涉及不同数量的圈，对应于不同的距离。

有时利奥觉得记住每个距离在池中有多少圈很困难，下面的表能够帮助他：

25码游泳池：
　　1个长度=25码（从墙到墙）
　　2个长度=50码
　　4个长度=100码
　　1/4英里=约500码=20个长度
　　1/2英里=约800码=32个长度
　　1英里=约1700码=68个长度
　　1.2英里=约2000码=80个长度
　　2.4英里=约4000码=160个长度

50米游泳池：
- 1个长度 = 50米
- 2个长度 = 100米
- 1/4英里 = 约400米 = 8个长度
- 1/2英里 = 约800米 = 16个长度
- 1英里 = 约1500米 = 30个长度
- 1.2英里 = 约2000米 = 40个长度
- 2.4英里 = 约4000米 = 80个长度

其中

$$1\text{米} = 1.0936\text{码}, \quad 1\text{码} = 0.9144\text{米}$$

你可以使用所有这些数字来计算利奥在训练中游的距离。

1. 如果利奥游100码，利奥游了多少个长度（圈数）？

利奥和他的同学在游泳课游了一个接力。每队5人，每人游4个25码池的长度。

2. 他们一共游了多少距离？

如果利奥进了校队，他将在一个50米泳池里游泳，这意味着每圈都要游更长的距离，使用上面米和码之间的转换。

3. 50米泳池的一个长度是多少码？ 50米泳池的一个长度比25码泳池的2个长度多了多少？

4. 如果利奥和他的4个朋友在50米泳池里进行同样的接力比赛，他们都游4个长度，他们游了多长距离？这距离是多少码？

13
曲棍球：角度

杰伊也是学校曲棍球队的队员，他在球队司职前锋，因此不时破门得分。杰伊很想成为球队中得分最高的球员，他需要更多练习才能变得更强。

数学对杰伊也有益处。正如杰伊已经发现的：曲棍球涉及很多角度。射门时，如果击球角度不对，是不会破门得分的。他知道为得到更多进球、变成更强的曲棍球运动员，去理解这些角度是值得的。试试你自己去理解曲棍球中的角度。

这里是曲棍球运动中常见的一些角度：

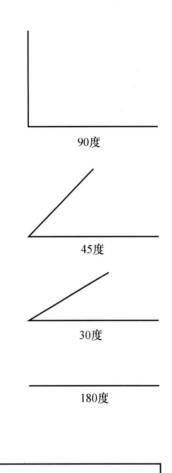

1. 你估算最左边的冰球到球门中间的角度是多少？

2. 你认为中间那个冰球到球门中间的角度是多少？

3. 右边的冰球呢？运动员能够在那个角度射入球门吗？

14
奥运会奖牌

奥运会每两年举办一次，夏季奥运会和冬季奥运会轮换举办。杰伊什么比赛都看，想看尽可能多的运动。

2012年，伦敦主办了夏季奥运会。比赛结束后，人们记录了每个国家赢得的奖牌数量，并相互比较。使用条形图，你可以轻松地比较不同国家在整体或不同运动的奖牌数量。图表提供了一个简单的方法来可视化奥运会上的表现。下面请填写你自己的图表。

2012年夏季奥运会获得奖牌最多的国家是：

美　国：46金，29银，29铜
中　国：38金，27银，27铜
俄罗斯：24金，26银，32铜
英　国：29金，17银，19铜
德　国：11金，19银，14铜

1. 每个国家获得奖牌的总数是多少？

完成奖牌数的条形图，以便你可以看到国家/地区间的比较。第一个国家已经完成了。

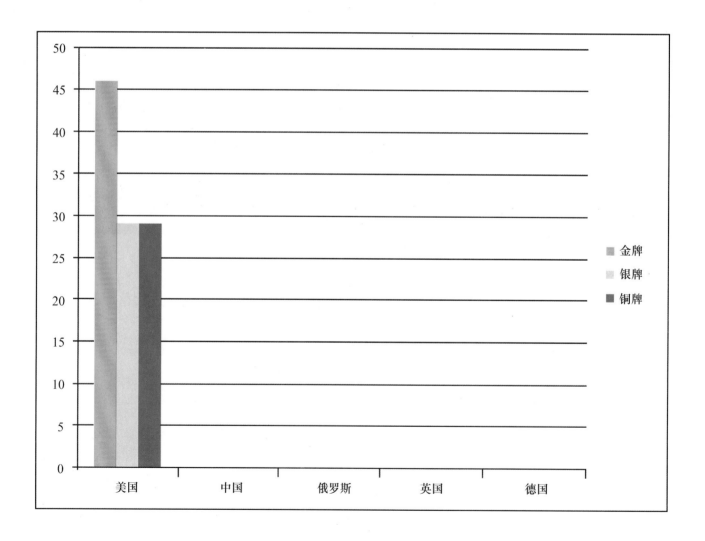

15
小　结

杰伊和家人、朋友们参加了很多运动。玩的时间越长,杰伊越是觉得数学在运动中有着重要的作用。从记录分数,到计算场地面积、找出最好的球员或球队,数学让运动更加有趣,并提升了比赛水平。下面,看看你是否记住了和杰伊一起学到的东西。

1. 在一次棒球比赛中，击球手有5次击球，其中有2次一垒安打。那么此次比赛他的打击率是多少？

2. 棒球赛季到目前为止，投手已经投了137又1/3局，其中自责分是39。那么他的防御率是多少？

3. 篮球队在23场比赛中赢了14场，这支队伍的胜负比是多少？

球队的胜率是多少？

4. 超级碗赛将在2030年举行，用罗马数字表示是什么？

5. 足球场的面积是6820平方码，宽度是62码，那么足球场有多长？

6. 某盘最后一局比赛中网球运动员的得分是30-0（她发球）。两个运动员在这局中平手，5-5。

她需要再得多少分才能赢得这局比赛？

获胜这局是否意味着选手赢得了这盘？为什么？

7. 高尔夫球场的一个4杆球洞，选手6次击球才将球送入洞。

那么这个洞他的分数是多少？

8. 一个25码的游泳池，游1英里是多少个长度？

在一个50米的游泳池呢？

参考答案

1.

1. 0.198 = 19.8%
   0.304 = 30.4%
   0.271 = 27.1%
   0.336 = 33.6%
2. 0.324
3. 62次
4. 是，他的打击率为0.218

2.

1. (2 × 9) ÷ 5 = 3.6
2. (89 × 9) ÷ 325 = 2.46
3. 26
4. 5.09, 4.89, 3.78, 2.55, 1.93

3.

1. 15
2. 2
3. 5
4. 4分，杰伊的球队赢了

4.

1. 73英寸；6' 1"
2. 74英寸（6' 2"）
3. 74英寸（6' 2"）
4. 9英寸

5.

1. 14-1
2. 14和15
3. 0.071/7.1%; 0.933/93.3%
4. A. 14-1
   B. 9-5
   C. 7-7
   D. 5-10
   E. 4-10
   F. 3-12
   G. 1-13

6.

1. 10分
2. 1次射门得分,1次2分转换,2次安全得分和1次触地附加或者2次2分转换,1次射门得分,1次安全得分和1次触地附加。其他满足要求的答案同样正确
3. 1次射门得分和1次安全得分,或者2次安全得分和1次触地附加。其他满足要求的答案同样正确
4. 是的,那样对手将比杰伊队多1分

7.

1. I, II, III, IV, V, VI, VII, VIII, IX, X
2. XLVI
3. LVII
4. MMXIII

8.

1. 6000平方码
2. 54000平方英尺
3. 50码(150英尺);27000平方英尺
4. 6码
5. 18英尺

9.

1. 15-40
2. 15-60, 这局结束
3. 关赢了这一盘
4. 比分变为6-5, 他们还需比赛至少一场
5. 是, 因为他将多赢出2分

10.

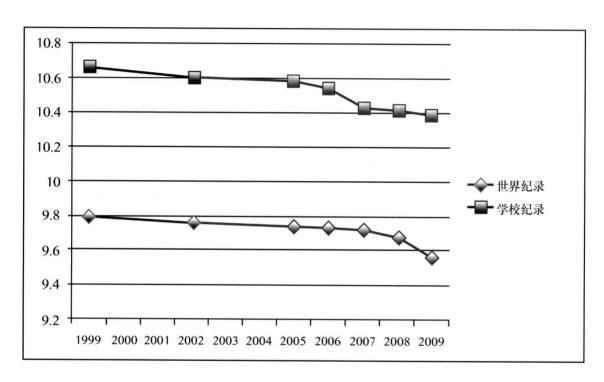

11.

1. 杆数低, 因为球容易打进去
2. 是
3. 5-杆洞
4. 0 + 1 - 2 = -1
5. - 1; - 2

12.

1. 4
2. 1/4 英里/500码
3. 50 × 1.0936 = 54.68码; 多了4.68码
4. 1000米; 1093.6码

13.

1. 30度
2. 45度
3. 90度; 不可能

14.

1. 美国 = 104, 中国 = 92, 俄罗斯 = 82, 英国 = 65, 德国 = 44

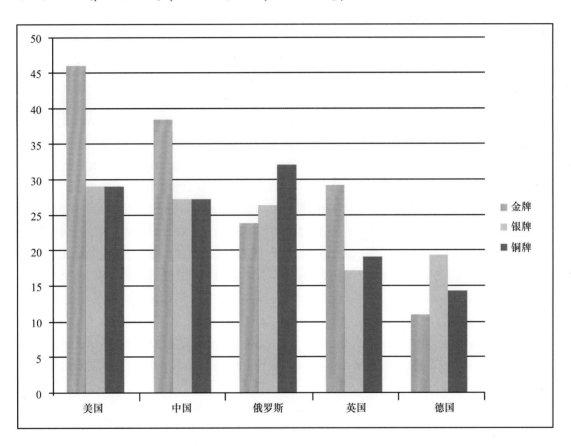

15.

1. 2次击中 ÷ 5次击球 = 0.400
2. (39 × 9) ÷ 137.33 = 2.56
3. 14- 9; 0.609/60.9%
4. LXIV
5. 6044 ÷ 62 = 97.48码
6. 至少2分; 不是, 她还没有赢得这盘, 因为她得多赢两局才行
7. 双柏忌/+2
8. 68个长度; 30个长度

INTRODUCTION

How would you define math? It's not as easy as you might think. We know math has to do with numbers. We often think of it as a part, if not the basis, for the sciences, especially natural science, engineering, and medicine. When we think of math, most of us imagine equations and blackboards, formulas and textbooks.

But math is actually far bigger than that. Think about examples like Polykleitos, the fifth-century Greek sculptor, who used math to sculpt the "perfect" male nude. Or remember Leonardo da Vinci? He used geometry—what he called "golden rectangles," rectangles whose dimensions were visually pleasing—to create his famous *Mona Lisa*.

Math and art? Yes, exactly! Mathematics is essential to disciplines as diverse as medicine and the fine arts. Counting, calculation, measurement, and the study of shapes and the motions of physical objects: all these are woven into music and games, science and architecture. In fact, math developed out of everyday necessity, as a way to talk about the world around us. Math gives us a way to perceive the real world—and then allows us to manipulate the world in practical ways.

For example, as soon as two people come together to build something, they need a language to talk about the materials they'll be working with and the object that they would like to build. Imagine trying to build something—anything—without a ruler, without any way of telling someone else a measurement, or even without being able to communicate what the thing will look like when it's done!

The truth is: We use math every day, even when we don't realize that we are. We use it when we go shopping, when we play sports, when we look at the clock, when we travel, when we run a business, and even when we cook. Whether we realize it or not, we use it in countless other ordinary activities as well. Math is pretty much a 24/7 activity!

And yet lots of us think we hate math. We imagine math as the practice of dusty, old college professors writing out calculations endlessly. We have this idea in our heads that math has nothing to do with real life, and we tell ourselves that it's something we don't need to worry about outside of math class, out there in the real world.

But here's the reality: Math helps us do better in many areas of life. Adults who don't understand basic math applications run into lots of problems. The Federal Reserve, for example, found that people who went bankrupt had an average of one and a half times more debt than their income—in other words, if they were making $24,000 per year, they had an average debt of $36,000. There's a basic subtraction problem there that should have told them they were in trouble long before they had to file for bankruptcy!

As an adult, your career—whatever it is—will depend in part on your ability to calculate mathematically. Without math skills, you won't be able to become a scientist or a nurse, an engineer or a computer specialist. You won't be able to get a business degree—or work as a waitress, a construction worker, or at a checkout counter.

Every kind of sport requires math too. From scoring to strategy, you need to understand math—so whether you want to watch a football game on television or become a first-class athlete yourself, math skills will improve your experience.

And then there's the world of computers. All businesses today—from farmers to factories, from restaurants to hair salons—have at least one computer. Gigabytes, data, spreadsheets, and programming all require math comprehension. Sure, there are a lot of automated math functions you can use on your computer, but you need to be able to understand how to use them, and you need to be able to understand the results.

This kind of math is a skill we realize we need only when we are in a situation where we are required to do a quick calculation. Then we sometimes end up scratching our heads, not quite sure how to apply the math we learned in school to the real-life scenario. The books in this series will give you practice applying math to real-life situations, so that you can be ahead of the game. They'll get you started—but to learn more, you'll have to pay attention in math class and do your homework. There's no way around that.

But for the rest of your life—pretty much 24/7—you'll be glad you did!

1
BASEBALL: BATTING AVERAGES

Jay loves sports. He plays several sports at his school, during the fall, winter, and spring. Plus, he follows as many sports as he can on TV, online, and in the newspapers. One of the sports Jay pays the most attention to is baseball. Jay also happens to love math, and baseball has a lot to do with math. Fans and professionals keep track of statistics, talk about how good players are in terms of numbers, and predict what will happen in games using math.

One of the numbers used in baseball is batting average. Batting average is a number that measures how often a batter gets a base hit. Batters who strike out all the time don't have very good batting averages. Batters who hit the ball often and get to base (singles, doubles, triples, or homeruns) will have a higher batting average. Only hits that get a batter to a base count. Look on the next page to find out how to calculate batting average.

Batting averages look like this: .320, .245, .222. They are really percentages, given in decimal form. To understand the percent form of batting averages, just move the decimal point two places to the right. That gives you the percent of times a player gets a base hit when he is at bat. So someone with a .320 batting average hits the ball 32% of the time. Someone with a .245 batting average hits the ball 24.5% of the time.

1. What are the percent forms of the following batting averages?

    .198 =
    .304 =
    .271 =
    .336 =

The formula for calculating batting average is:

$$\text{batting average} = \text{number of hits} \div \text{number of at bats}$$

If a player has 30 at bats, and he gets a base hit 9 times, his batting average would be:

$$9 \text{ hits} \div 30 \text{ at bats} = 0.300$$

There are always three numbers in a batting average, so you always have to write your calculations of batting average up to the thousandths place (the third place to the right of the decimal).

2. What is the batting average last season for Jay's favorite player if he has gotten 175 hits in 540 at bats?

You can also figure out how many hits a batter has gotten if you know his batting average and how many times he has been at bat. You just need to rearrange the equation a little:

$$\text{number of hits} = \text{batting average} \times \text{number of at bats}$$

3. If a player has a batting average of .322 and has been at bat 193 times, how many hits has he gotten?

In modern-day baseball, players almost never have batting averages that are above .400. Batting averages below .200 are considered poor.

4. If a player has 77 hits in 353 at bats, does he fall within this range? What is his batting average?

2
BASEBALL: EARNED RUN AVERAGES

While watching baseball on TV, Jay particularly likes to pay attention to the pitchers. He used to be a pitcher when he played Little League baseball when he was younger, and he likes to be the pitcher when he is playing baseball with his friends.

While batting average tells Jay how good batters are, Earned Run Average (ERA) tells him how good pitchers are. ERA is a measure of how many runs a pitcher gives up to the other team. The lower an ERA is, the better the pitcher is, because it means a pitcher didn't allow the other team to score many runs. You'll find out how to use and calculate ERA on the next page.

To calculate a pitcher's ERA for one game, add up all the innings he pitched. Pitchers don't usually pitch for the whole game, so he may have pitched for 2, 5, or 8 innings. Include parts of innings. If a player pitched one-third of an inning, add .33. If he pitched 2/3 of an inning, add .67.

Now add up all the runs the other team got in those innings, which are the pitcher's earned runs. Multiply the earned runs by 9.

Finally, divide that number by the number of innings the pitcher pitched. Make sure there are only two numbers after the decimal point. The whole equation looks like this:

$$\text{ERA} = (\text{earned runs} \times 9) \div \text{innings pitched}$$

1. What would be the ERA of a pitcher who pitched 5 innings in a game, and gave up 2 runs?

You can do the same thing to find the ERA of pitchers over a whole season, or over a player's career. You just use bigger numbers.

2. Last season, a pitcher pitched in 325 innings. He gave up 89 runs. What was his ERA?

Like with batting average, you can rearrange the ERA equation to find missing information. For example, if you know a pitcher's ERA and how many innings he played, you can find out how about many runs he gave up.

3. How many runs did a pitcher give up last month if he played 56.67 innings, and had an ERA of 4.20? The number of earned runs has to be a whole number.

   Earned runs = (ERA x innings pitched) ÷ 9
   Earned runs = (4.20 x 56.67) ÷ 9
   Earned runs =

An ERA under 2.00 is amazing. An ERA over 5.00 is poor and shows the pitcher is struggling.

4. Here are several ERAs: 3.78, 4.89, 1.93, 2.55, 5.09. Order the ERAs from least talented to most talented:

3
BASEBALL: SCORING

When Jay goes to baseball games, he likes to fill out a scorecard. He can see the score on the scoreboard, but he wants to understand how the teams get their scores.

A lot of numbers are used to score baseball. There are 3 strikes in an out and 4 balls in a walk. There are 3 outs in an inning, and 9 innings in a game. There are singles, doubles, triples, and homeruns. Hitting a single gets a player to first base. If any other players are on base, they advance one base as well. Hitting a double gets the player to second base and advances other players one or two bases, depending on how far they can get. Hitting a triple gets the player to third base and advances other players at least one base. And hitting a homerun gets the player all the way home, as well as any other players on base!

Jay finds filing out a scorecard makes it easier to keep track of all those numbers! On the next page, you'll figure out how to score a game yourself.

Here is an overview of everything that happened to Jay's favorite team in the most recent game Jay watched:

1st inning: Player 1—3 strikes; Player 2—single; Player 3—3 strikes; Player 4—double (Player 2 advances two bases); Player 5—single (Player 2 advances to home, Player 4 out at second base)

2nd inning: Player 1—fly out; Player 2—4 balls; Player 3—3 strikes; Player 4—3 strikes

3rd inning: Player 1—3 strikes; Player 2—fly out; Player 3—home run; Player 4—fly out

4th inning: Player 1—4 balls; Player 2—single (Player 1 advances one base); Player 3—single (Player 1 advances 1 base, Player 2 advances 1 base); Player 4—double (Player 1 advances home, Player 2 out at third base, Player 3 out at second base); Player 5—3 strikes

5th inning: Player 1—single; Player 2—fly out; Player 3—3 strikes: Player 4—3 strikes

6th inning: Player 1—3 strikes; Player 2—double; Player 3—3 strikes; Player 4—single (Player 2 advances to third base); Player 5—fly out

7th inning: Player 1—3 strikes; Player 2—3 strikes; Player 3—3 strikes

8th inning: Player 1-fly out; Player 2—fly out; Player 3—double; Player 4—3 strikes

9th inning: Player 1—3 strikes; Player 2—home run; Player 3—single; Player 4—fly out

1. How many strikeouts were there?

2. How many walks were there?

3. How many times did a player safely get to third base or farther?

4. What was the final score? If the other team had 3 points, did Jay's team win or lose?

4
BASKETBALL: AVERAGE HEIGHT

Jay also likes basketball. He likes to watch it on TV and go to games when he can. He plays basketball with his friends whenever he can.

Jay has never played basketball at school, though, mostly because he isn't tall enough. Basketball players tend to be really tall, so they can more easily reach the basket. You don't have to be tall to enjoy playing basketball, or even be good at it, but coaches usually want tall players.

College and professional basketball players—both women and men—are often really tall. Calculate different height averages on the next page, including mean, median, and mode.

Here are the heights on the 2012 roster for the Seattle Storm, two-time WNBA champions:

6'2" (74 inches)
5'9" (69 inches)
5'10" (70 inches)
6'6" (78 inches)
6'3" (75 inches)
6'2" (74 inches)
5'11" (71 inches)
6'2" (74 inches)
6'2" (74 inches)
6'4" (76 inches)
5'11" (71 inches)

The average most people are used to talking about is called the mean. You need to add up all the heights and then divide by the number of heights you added. It's easiest to calculate the mean just in inches, and then convert to feet and inches.

1. What is the mean of their heights in inches? Round to the nearest whole inch. What is

that in feet and inches?

The median is another average. Arrange the heights in order from least to greatest. Then cross off the shortest and tallest players. Keep doing that until you arrive at one number in the middle. If there are two numbers, you can take the mean of them to find the number that is exactly in the middle.

2. What is the median for these heights?

A third kind of average is called the mode. The mode is the height that occurs the most.

3. What is the mode?

4. If the average height for women in the United States is around 5'4", how many inches taller is the average women on the Seattle Storm?

5
BASKETBALL: WIN-LOSS

During basketball season, Jay likes to follow every team in the college and professional leagues. He checks online for the teams' **standings** every week. The number he looks at is the win-loss number. When he sees a number like 9–3, he knows that team won 9 games and lost 3. They played a total of 12 games.

Another number that goes along with win-loss is win percentage. Win percentage is a measure of how many games a team has won compared to how many they have played. Then you can take a look at every team's win percentage, and you can tell which teams have played the best during a season so far. The next page will help you practice understanding win-loss and win percentages.

Jay takes a look at the win-loss numbers for several basketball teams midway through the season. He sees:

4–10
14–1
5–10
9–5
7–7
3–12
1–13

1. Which team has a better win-loss record, the team with 1–13 or the team with 14–1?

2. How many games have those two teams played?

You can see some of the teams have played different numbers of games, so comparing them to each other gets a little tricky. To figure out just how well each team has done in comparison to the other teams, you'll need to find the win percentage. The formula is:

win percentage = wins ÷ total games played.

In basketball, win percentage is usually shown as a decimal number, with three decimal places. To convert to a normal percent, just move the decimal point two places to the left.

3. What is the win percentage for the 1–13 team and the 14–1 team? Give your answers in both decimal and percent form.

4. Now you can order the teams from first to last, starting with the team with the highest win percentage (A) and ending with the team with the lowest (G).

   A.
   B.
   C.
   D.
   E.
   F.
   G.

6 FOOTBALL: SCORING

Another sport Jay likes is football. He watches games on TV, plays for fun with his friends on the weekend, and even plays on his school's team. He dreams about becoming a professional football player in the future.

Football, like other sports, has a lot of numbers and math. Anyone who watches or plays football has to learn the many different ways of scoring points. Every game is different, even if the scores end up being the same. Find out on the next page how you can combine points to get scores.

Here are the ways players score points in a football game:

a touchdown = 6 points

an extra point after a touchdown = 1 point

a 2-point conversion = 2 points

a field goal = 3 points

a safety = 2 points

In the last football game Jay played in, his high school team won with 34 points. The other team scored 29 points.

1. Jay's team scored four touchdowns. How many points did they score in other ways besides touchdowns?

2. What is one way the team could have scored those points using all four other ways of scoring?

3. What is one way the losing team could have scored 29 points if they also had four touchdowns?

4. If the other team had gotten one more touchdown, would they have won the game? Why or why not?

7
FOOTBALL: ROMAN NUMERALS

The Super Bowl is one of Jay's favorite days. He and his family all watch the Super Bowl together. They make special food and wear their team's jerseys.

In 2013, they watched the forty-seventh annual Super Bowl. However, the game isn't advertised using the normal numbers we use today in math class. Instead, the year of the Super Bowl is given in Roman numerals, so the 2013 game was Super Bowl XLVII. Jay's younger sister was a little confused about how to read Roman numerals, so Jay taught her how during the game. You'll learn too, on the next page.

In Roman numerals, different letters stand for certain numbers.

I = 1
V = 5
X = 10
L = 50
C = 100
D = 500
M = 1,000

You form numbers by adding different letters together. Two would just be two 1s, or II. Three would be three 1s, or III. But four is a little different. You don't want to list four IIIIs in a row. Instead, you write IV, which means "one less than five." You do the same thing for ten, but using an X instead of a V.

1. What is one through ten in Roman numerals?

For the forty-seventh Super Bowl, the Roman numeral is XLVII. You can divide that into different numbers you add together. Divide it by places: first look at the tens place, then the ones place.

The XL is the tens place. It really means "ten less than fifty." You might have thought XL would mean 60, because ten and fifty added together is 60. But every time you see a letter that has less **value** than the letter that comes after it, you subtract it instead of add it. If it were 60, it would read LX, and the ten would come after the fifty.

After XL is VII. That's just 5 plus 2, or 7. Add the XL in the tens' place and the VII in the ones' place together and you get 47.

2. What was Super Bowl forty-six in Roman numerals?

3. In 2023, it will be Super Bowl fifty-seven. What will that be in Roman numerals?

You can also give larger numbers in Roman numerals, like the year of the forty-seventh Super Bowl—2013.

4. What is 2013 in Roman numerals?

8
SOCCER: FIELD AREA

Jay's friend Ananya loves to play soccer. She is one of the star players on her school team, and she wants to become a professional soccer player someday.

Ananya and Jay's school is building a new soccer field because their old one was flooded and was really old. The new soccer field has to be built using exact measurements. The **dimensions** of the field are important. Soccer fields at every school and in professional games are around the same size so players get used to how big they are and what to expect. A team won't have an unfair advantage when other teams come to play on their field and aren't used to the size.

The next page will take you through soccer field dimensions, and will help you calculate the area of the field.

The new soccer field, at the school will be 100 yards long and 60 yards wide. (Remember, 1 yard equals 3 feet.)

You already have enough information to calculate the area of the field. The area of a rectangle is:

$$area = length \times width$$

1. What is the area of the field in square yards?

Now find the area in square feet. First, convert the yards into feet by multiplying each dimension by 3. Then plug those numbers into the area equation.

2. What is the area of the field in square feet?

The school also has to draw lots of lines on the field and set up the goals. The midfield line, for example, divides the field in half lengthwise, and shows which side the teams are defending.

3. How far down the length of the field (in feet) should the midfield line be drawn?

   What is the area of each half of the field in square feet?

On either end of the field, in front of the goals, are the goal areas. Players have to kick the ball from this area to score a goal.

Goal areas are 120 square yards in area. If one side of the goal area is 20 yards long, you can find the missing piece of information (the width) by rearranging the area equation.

The new equation would be:

$$width = area \div length$$

4. So, the width = 120 square yards ÷ 20 yards =

5. What is the width of the goal area in feet?

9
TENNIS: SCORING

Jay's friend Kwan plays tennis. Sometimes Jay goes and watches his tennis matches and cheers his friend on. Kwan is pretty good—he wins a lot of the time.

At first, figuring out how tennis scores work was a little confusing for Jay, because tennis doesn't just count points from 1 on up. Instead of 1, 2, 3, 4, and so on, tennis uses a different scoring system. Jay quickly figured it out, and you can too.

Here is how tennis is scored:

$$0 \text{ points} = \text{love}$$
$$1 \text{ point} = 15 \text{ game points}$$
$$2 \text{ points} = 30 \text{ game points}$$
$$3 \text{ points} = 40 \text{ game points}$$
$$4 \text{ points} = 60 \text{ game points/game over}$$

And here are some of the rules of scoring one game:

- You must earn a minimum of 4 points to win a game.
- You must win by 2 points.
- If you are not ahead by 2 points by the time you get 40 game points, the next point you score is called advantage. If you get another point, you win the game.
- The score is announced server first.

In one game Jay watches Kwan play, Kwan has scored 1 point and his opponent has scored 3 points. Kwan is the one serving next.

1. What is the score in tennis terms?

Kwan's opponent scores the next point.

2. What is the score? Is the game over?

Kwan and Jay aren't too worried, though. Losing this game doesn't mean Kwan has lost the whole match. In tennis, you need to win 6 games to win one set. You also have to win by 2 games in a set.

As the game goes on, Kwan ends up winning 5 games. His opponent wins 4.

3. What will happen in this set if Kwan wins the next game?

4. What will happen in this set if Kwan's opponent wins the next game?

Finally, tennis players usually play 3 sets in the whole match. The winner has to win two out of the three sets.

Kwan wins one set and his opponent wins one set. They have to play one more to decide who will win the match. With one game to go, the final set is scored 5–4 in favor of Kwan.

5. Will Kwan win the match if he wins this set? Why or why not?

10 TRACK: RECORDS

Once in a while, Jay goes to a track meet at his school. He has a few friends on the team, and he likes to cheer them on.

He wonders what the world records are for track. His friends are fast, but he knows that the men and women who train to run can go a lot faster! When he gets home after a track meet, he looks up the world records for track. He can see they are really fast, but he isn't sure exactly how the times compare to his friends' times.

Putting track time **data** on a line graph could help Jay compare the times. You can do the same thing on the next page.

Jay decides to compare records for the men's 100-meter dash, the shortest track event. Here's the information he finds for world records in the last few years:

2009 world record: 9.58 seconds
2008 world record: 9.69 seconds
2007 world record: 9.74 seconds
2006 word record: 9.762 seconds
2005 world record: 9.768 seconds
2002 world record: 9.78 seconds
1999 world record: 9.79 seconds

And here is the data for his school for the same years:

2009 school record: 10.39
2008 school record: 10.41
2007 school record: 10.42
2006 school record: 10.56
2005 school record: 10.59
2002 school record: 10.60
1999 school record: 10.67

Now fill out the rest of this line graph, with two lines: one for the world record and one for the school record.

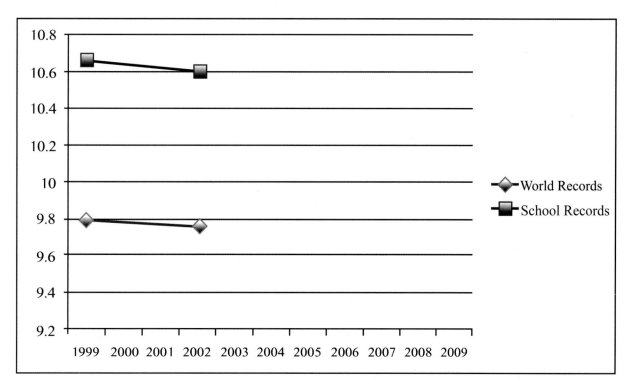

11
GOLF: PAR

Jay is flipping through the channels on his TV when he comes across golf. He hasn't really watched much golf before, but he decides he wants to learn more about how golf is played and scored.

He finds that like other sports, golf has plenty of numbers involved. In golf, players go through 18 different holes. Each hole has a par, which is the number of shots it should take to get the golf ball in the hole. Harder holes have higher pars, while easier holes have lower pars. Players win by getting the ball in the holes in as few strokes as possible.

Jay figures out the rules pretty quickly, and can tell who the best players are from their scores on individual holes. See if you can do the same on the next page.

Par is based on how difficult the hole is. Holes with higher pars are generally really long. They also may have sharp turns or hills.

1. Would a hole that is 200 yards long and straight have a higher or lower par than a hole that is 320 yards long with a curve in the middle? Why?

Golf courses often have 18 holes that add up to 72, though not always.

2. Does a course with four par-3 holes, ten par-4 holes, and four par-5 holes add up to 72?

3. Which are the most difficult holes on this golf course?

Golf has different terms for how many strokes it takes to get the ball in the holes, besides par.

double eagle: 3 strokes under par, or –3
eagle: 2 strokes under par, or –2
birdie: 1 stroke under par, or –1
par: par, or 0

bogey: one stroke over par, or +1
double bogey: two strokes over par, or +2
triple bogey: three strokes over par, or +3

The best players get birdies, eagles, or double eagles.

To calculate a player's score, just add up the numbers associated with the number of shots above or below par it took for the player to get the ball in the hole.

The player Jay is watching gets a par, a bogey, and an eagle on the first 3 holes.

4. What is her overall score?

The next hole is a par-5 hole. She takes 4 strokes to get the ball in the hole.

5. What is her score on this hole? What is her overall score now that she has played another hole?

12
SWIMMING: LAPS AND DISTANCE

Jay's younger brother Leo takes swimming lessons. He wants to be on the school team eventually, but he has to practice more first. Right now, he takes lessons in a 25-yard pool. If he joins the swim team, though, he'll swim in a 50-meter pool, which is a little bigger.

Every lap Leo swims in either pool is a certain distance. Swim competitions involve different numbers of laps, which **correspond** to different distances.

Sometimes Leo has trouble remembering how many laps in the pool each distance is. These tables will help him:

25-Yard Pool
1 length = 25 yards (from wall to wall)
2 lengths = 50 yards
4 lengths = 100 yards
¼ mile = about 500 yards = 20 lengths

½ mile = about 800 yards = 32 lengths
1 mile = about 1700 yards = 68 lengths
1.2 miles = about 2000 yards = 80 lengths
2.4 miles = about 4000 yards = 160 lengths

50-Meter Pool
1 length = 50 meters
2 lengths = 100 meters
¼ mile = about 400 meters = 8 lengths
½ mile = about 800 meters = 16 lengths
1 mile = about 1500 meters = 30 lengths
1.2 miles = about 2000 meters = 40 lengths
2.4 miles = about 4000 meters = 80 lengths

1 meter = 1.0936 yards/ 1 yard = 0.9144 meters

You can use all these numbers to figure out how far Leo swims during his practices.

1. How many lengths of the pool (laps) will Leo swim if he swims 100 yards?

Leo and his classmates are doing a relay in swim class. There are 5 people per team and each of them swims 4 lengths of the 25-yard pool.

2. What distance do they all swim together?

If Leo makes it onto the school swim team, he will be swimming in a 50-meter pool, which means he will be swimming longer distances for each lap. Use the conversions between meters and yards above.

3. How long is one length of the 50-meter pool in yards? How much longer is one length of the 50-meter pool than two lengths of the 25-yard pool?

4. If Leo and his 4 friends were to do the same relay race in the 50-meter pool, what distance would they have swum if they all swam four lengths? What is that distance in yards?

13 HOCKEY: ANGLES

Jay is also on his school's hockey team. He has an **offensive** position, so he sometimes scores goals. He isn't the highest-scoring player on his team, but he'd like to be. He needs to practice more to get even better.

Math can also help Jay get better. Hockey involves a lot of angles, as Jay has found out. If he doesn't shoot the puck at the right angle when he's trying to score a goal, he won't make it. He knows it's worth it to understand angles in order to score more goals and become a better hockey player. Try out your own understanding of angles in hockey.

Here are some common angles you'll find in hockey:

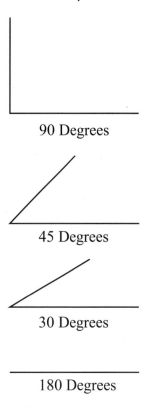

90 Degrees

45 Degrees

30 Degrees

180 Degrees

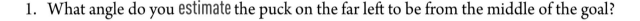

1. What angle do you estimate the puck on the far left to be from the middle of the goal?

2. What angle do you think the middle puck is from the middle of the goal?

3. And what about the puck on the right? Will the player be able to shoot it into the goal at that angle?

14
OLYMPIC MEDALS

The Olympics happen every two years, rotating between the summer and winter Olympics. Jay watches all of them, and tries to see as many sports as possible.

In 2012, London hosted the summer Olympics. After the games, people tallied up the number of medals each country had won, and compared them to each other. You can use bar graphs to easily compare different countries overall and in different sports. Graphs offer an easy way to visualize what is going on in the Olympics. Turn to the next page to fill in your own graph.

56

The countries that got the most medals in the 2012 Summer Olympics were:

United States: 46 gold, 29 silver, 29 bronze
China: 38 gold, 27 silver, 27 bronze
Russia: 24 gold, 26 silver, 32 bronze
Great Britain: 29 gold, 17 silver, 19 bronze
Germany: 11 gold, 19 silver, 14 bronze

1. What was the total number of medals each country won?

   Finish out this bar graph of medal scores, so that you can see how each country compares. The first country has been done for you.

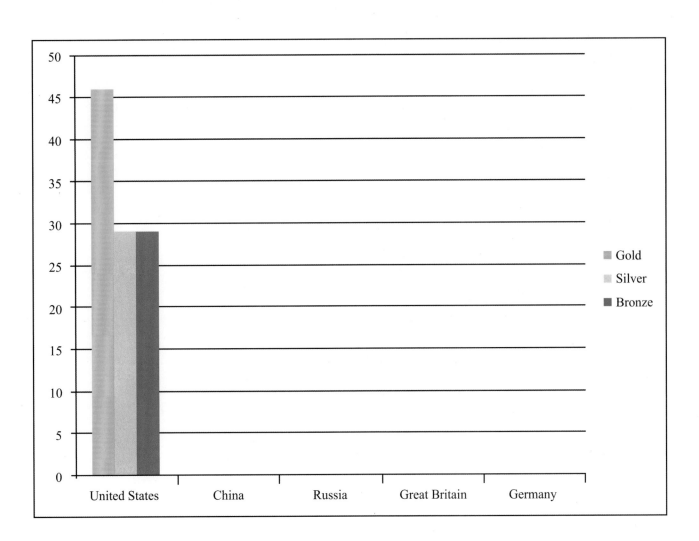

15 PUTTING IT ALL TOGETHER

Jay and his family and friends play a lot of sports. The longer he plays, the more Jay understands that math is a big part of sports. From keeping track of scores, to figuring out the area of a field, to figuring out who is the best player or team, math makes sports more fun and improves the game. On the next page, see if you can remember some of what you have learned along with Jay.

1. In one baseball game, a batter has 5 at bats. He gets a base hit twice in those 5 at bats. What is his batting average for the game?

2. A pitcher has pitched 137 ⅓ innings this baseball season so far. He has given up 39 runs during those innings. What is his ERA?

3. A basketball team wins 14 games out of the 23 games it plays. What is the team's win-loss number?

    What is the team's win percentage?

4. What Super Bowl will be played in the year 2030, in Roman numerals?

5. A soccer field's area is 6820 square yards and it is 62 yards wide. How long is the soccer field?

6. A tennis player's score during the last game of a set is 30–love (she is serving). The two players are tied in the set, 5–5 .

   How many more points does she need to win the game?

   Does winning the game mean the player wins the set? Why or why not?

7. On a par-4 hole at the golf course, a player takes 6 strokes to get the ball in the hole.

   What is his score for the hole?

8. In a 25-yard pool, how many lengths would it take to swim a mile?

   What about in a 50-meter pool?

Answers

1.

1. .198 = 19.8%
   .304 = 30.4%
   .271 = 27.1%
   .336 = 33.6%
2. .324
3. 62 hits
4. Yes, he has a batting average of .218.

2.

1. (2 x 9) ÷ 5 = 3.6
2. (89 x 9) ÷ 325 = 2.46
3. 62
4. 5.09, 4,89, 3.78, 2.55, 1.93

3.

1. 15
2. 2
3. 5
4. They had 4 points, and won the game.

4.

1. 73 inches; 6'1"
2. 74 inches/6'2"
3. 74 inches/6'2"
4. 9 inches

5.

1. 14–1
2. 14 and 15
3. $0.071/7.1\%$; $0.933/93.3\%$
4. A. 14–1
   B. 9–5
   C. 7–7
   D. 5–10
   E. 4–10
   F. 3–12
   G. 1–13

6.

1. 10 points
2. One field goal, a 2-point conversion, 2 safeties, and an extra touchdown point OR 2 2-point conversions, a field goal, a safety, and an extra touchdown point. There are more answers as well.
3. One field goal and a safety, OR two safeties and a touchdown point. There are many more answers.
4. Yes, they would have had 1 more point than Jay's team.

7.

1. I, II, III, IV, V, VI, VII, VIII, IX, X
2. XLVI
3. LVII
4. MMXIII

8.

1. 6000 square yards
2. 54,000 square feet
3. 50 yards/150 feet; 27,000 square feet
4. 6 yards
5. 18 feet

9.

1. 15–40
2. 15–game, yes the game is over.
3. Kwan wins the set.
4. The score will be 6-5, and they will need to play at least one more game.
5. Yes, because he will have won it by 2 points.

10.

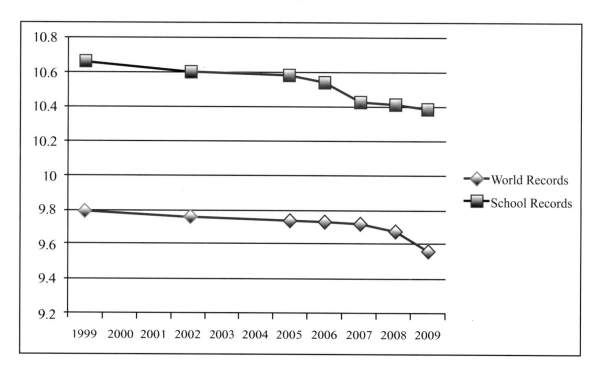

11.

1. Lower par, because it is easier to get the ball in the hole.
2. Yes
3. The par-5 holes
4. 0 + 1 − 2 = −1
5. −1; −2

12.

1. 4
2. ¼ mile/500 yards
3. 50 x 1.0936 = 54.68 yards; 4.68 yards longer
4. 1000 meters; 1,093.6 yards

13.

1. 30 degrees
2. 45 degrees
3. 90 degrees; no

14.

1. US = 104, China = 92, Russia = 82, Great Britain = 65, Germany = 44

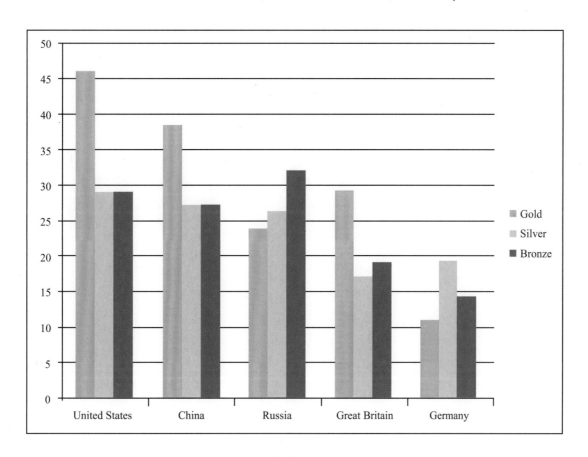

15.

1. 2 hits ÷ 5 at bats = .400
2. (39 x 9) ÷ 137.33 = 2.56
3. 14–9; 0.609/60.9%
4. LXIV
5. 6044 ÷ 62 = 97.48 yards
6. 2 more; no, she doesn't win the set, because she has to win by two games.
7. Double bogey/+2
8. 68 lengths; 30 lengths

本书由中国科学院数学与系统科学研究院资助出版

数学 24/7

时尚中的数学

〔美〕雷·西蒙斯 著

王晓欢 译

科学出版社

北京

图字：01-2015-5621号

内 容 简 介

时尚中的数学是"数学生活"系列之一，内容涉及衣服的尺寸及不同计量单位间的转换，促销折扣及优惠的计算，自己制作衣服时所用的方案设计、成本计算以及后期的店面和在线销售，并计算利润，让青少年在学校学到的数学知识应用到与时尚有关的多个方面，让青少年进一步了解数学在日常生活中是如何运用的。

本书适合作为中小学生的课外辅导书，也可作为中小学生的兴趣读物。

Copyright © 2014 by Mason Crest, an imprint of National Highlights, Inc. All rights reserved. No part of this publication may be reproduced or transmitted in any form or by any means, electronic or mechanical, including photocopying, recording, taping or any information storage and retrieval system, without permission from the publisher.
The simplified Chinese translation rights arranged through Rightol Media.
（本书中文简体版权经由锐拓传媒取得Email:copyright@rightol.com）

图书在版编目（CIP）数据

时尚中的数学/（美）雷·西蒙斯（Rae Simons）著；王晓欢译.—北京:科学出版社,2018.5
（数学生活）
书名原文: Fashion Math
ISBN 978-7-03-056747-5

Ⅰ.①时… Ⅱ.①雷… ②王… Ⅲ.①数学-青少年读物 Ⅳ.①O1-49

中国版本图书馆CIP数据核字（2018）第046671号

责任编辑:胡庆家 / 责任校对:邹慧卿
责任印制:肖 兴 / 封面设计:陈 敬

科学出版社 出版
北京东黄城根北街16号
邮政编码：100717
http://www.sciencep.com

北京汇瑞嘉合文化发展有限公司 印刷
科学出版社发行 各地新华书店经销
*
2018年5月第 一 版　　开本:889×1194 1/16
2018年5月第一次印刷　　印张:4 1/4
字数:70 000

定价：98.00元（含2册）
（如有印装质量问题，我社负责调换）

引　　言

你会如何定义数学？它也许不是你想象的那样简单。我们都知道数学和数字有关。我们常常认为它是科学，尤其是自然科学、工程和医药学的一部分，甚至是基础部分。谈及数学，大多数人会想到方程和黑板、公式和课本。

但其实数学远不止这些。例如，在公元前5世纪，古希腊雕刻家波留克列特斯曾经用数学雕刻出了"完美"的人体像。又例如，还记得列昂纳多·达·芬奇吗？他曾使用有着赏心悦目的尺寸的几何矩形——他称之为"黄金矩形"，创作出了著名的画作——蒙娜丽莎。

数学和艺术？是的！数学对包括医药和美术在内的诸多学科都至关重要。计数、计算、测量、对图形和物理运动的研究，这些都被融入到音乐与游戏、科学与建筑之中。事实上，作为一种描述我们周围世界的方式，数学形成于日常生活的需要。数学给我们提供了一种去理解真实世界的方法——继而用切实可行的途径来控制世界。

例如，当两个人合作建造一样东西时，他们肯定需要一种语言来讨论将要使用的材料和要建造的对象。但如果他们建造的过程中没有用到一个标尺，也不用任何方式告诉对方尺寸，甚至他们不能互相交流，那他们建造出来的东西会是什么样的呢？

事实上，即便没有察觉到，但我们确实每天都在使用数学。当我们购物、运动、查看时间、外出旅行、出差办事，甚至烹饪时都用到了数学。无论有没有意识到，我们在数不清的日常活动中用着数学。数学几乎每时每刻都在发生。

很多人都觉得自己讨厌数学。在我们的想象中，数学就是枯燥乏味的老教授做着无穷无尽的计算。我们会认为数学和实际生活没有关系；离开了数学课堂，在真实世界里我们再不用考虑与数学有关的事情了。

然而事实却是数学使我们生活各方面变得更好。不懂得基本的数学应用的人会遇到很多问题。例如，美联储发现，那些破产的人的负债是他们所得收入的1.5倍左右——换句话说，假设他们年收入是24000美元，那么平均负债是36000美元。懂得基本的减法，会使他们提前意识到风险从而避免破产。

作为一个成年人，无论你的职业是什么，都会或多或少地依赖于你的数学计算能力。没有数学技巧，你就无法成为科学家、护士、工程师或者计算机专家，就无法得到商学院学位，就无法成为一名服务生、一位建造师或收银员。

体育运动也需要数学。从得分到战术，都需要你理解数学——所以无论你是想在电视上看一场足球比赛，还是想在赛场上成为一流的运动员，数学技巧都会给你带来更好的体验。

还有计算机的使用。从农庄到工厂、从餐馆到理发店，如今所有的商家都至少拥有一台电脑。千兆字节、数据、电子表格、程序设计，这些都要求你对数学有一定的理解能力。当然，电脑会提供很多自动运算的数学函数，但你还得知道如何使用这些函数，你得理解电脑运行结果的含义。

这类数学是一种技能，但我们总是在需要做快速计算时才会意识到自己需要这种技能。于是，有时我们会抓耳挠腮，不知道如何将学校里学的数学应用在实际生活中。这套丛书将助你一马当先，让你提前练习数学在各种生活情境里的运用。这套丛书将会带你入门——但如果想掌握更多，你必须专心上数学课，认真完成作业，除此之外再无捷径。

但是，付出的这些努力会在之后的生活里——几乎每时每刻（24/7）——让你受益匪浅！

目 录

引言
1. 衣服尺码与数学   1
2. 尺码转换与数学   3
3. 鞋子尺码与数学   5
4. 消费中的数学   7
5. 促销中的数学   9
6. 在线购物中的数学   11
7. 量体裁衣中的数学   13
8. 处理旧衣服中的数学   15
9. 裁剪圆形中的数学   17
10. 裁剪矩形中的数学   19
11. 购买面料中的数学   21
12. 服装成本中的数学   23
13. 销售提成中的数学   25
14. 在线销售中的数学   27
15. 小结   29
参考答案   32

Contents

INTRODUCTION
1. SIZE MATH	37
2. MORE SIZE MATH: CONVERSIONS	39
3. SHOE SIZE	40
4. SPENDING MONEY	42
5. SALES MATH	43
6. SHOPPING ONLINE	45
7. TAILORING	46
8. RESELLING OLD CLOTHES	48
9. SEWING: CIRCUMFERENCE AND DIAMETER	49
10. SEWING: RECTANGULAR DIMENSIONS	51
11. CHOOSING FABRIC	52
12. FASHION BUSINESS: FIGURING OUT PRICES	54
13. SELLING ON COMMISSION	55
14. SELLING ONLINE	56
15. PUTTING IT ALL TOGETHER	58
ANSWERS	59

1
衣服尺码与数学

莫妮卡和朋友们约定一起去购物,她们打算去购物中心逛一天。莫妮卡打算买一条牛仔裤,她的朋友杰森打算买几件T恤,珍妮为了参加校园舞会打算买一条漂亮的裙子。

莫妮卡非常喜欢和朋友们一起去购物,但是偶尔莫妮卡会因为没办法通过观察确认哪件衣服适合自己,然后花了好几个小时,试了一件又一件衣服,最终却没有一件合适的,这个时候莫妮卡简直沮丧极了。

聪明的莫妮卡想到了一个好办法,如果知道自己衣服的平均尺码,就能更容易找到适合自己的衣服。下面就让莫妮卡教你如何计算自己的衣服尺码吧!

莫妮卡把最近买的几条牛仔裤拿了出来，看看它们的尺码是多少，其中一条是9码，两条是11码，一条是13码，一条是15码。

莫妮卡计算平均尺码的方法就是：将所有牛仔裤的尺码加起来，然后再除以牛仔裤的条数。

1. 那么，请你和莫妮卡一起来算算她的平均尺码是多少呢？

当你用上面这个数学方法来计算平均尺码时，可能得到的是一个小数，但是显然，商场里的衣服尺码是没有小数的。以莫妮卡为例，由于已有裤子的尺码都是奇数，所以她的平均尺码就是最接近那个平均值的奇数。

2. 以莫妮卡为例，请你说出最接近平均值的牛仔裤尺码？

通过这个办法，聪明的莫妮卡知道了以后购物时会先试哪个尺码的衣服。

同样地，你也可以帮助朋友算出他们衣服的平均尺码。

杰森有10件T恤，其中3件S码，5件M码，1件L码和1件XL码。显然你不能用S，M，L，XL这些字母来计算平均值，办法就是可以为每个字母假定一个整数与之对应：

$$S = 1$$
$$M = 2$$
$$L = 3$$
$$XL = 4$$

现在，你来算算杰森T恤的平均尺码吧。

3. 杰森T恤的平均尺码是多少？你可能还得将计算出的平均值四舍五入。

同样的办法，请你再来算算珍妮裙子的平均尺码吧！珍妮有5条裙子，其中一条S码，一条M码，3条L码。

4. 那么珍妮裙子的平均尺码是多少呢？

2

尺码转换与数学

莫妮卡和朋友们又进了另一家服装店,她选中三条牛仔裤想试穿一下,问题是,她看不懂这三条牛仔裤的尺码。正常情况下,莫妮卡穿美码11码的牛仔裤,偶尔也穿13码的。但这三条牛仔裤并不是美码的,而是两条女码的,一条欧码的。

难道要一条一条试穿吗?莫妮卡很发愁。还好,售货员帮了她大忙。售货员告诉她美码、欧码和女码之间换算的方法。通常情况下,女码裤子尺码都略微偏小,欧码裤子尺码标注的是腰围的大小。接下来就让我们和莫妮卡一起学学如何换算吧!

首先，售货员告诉莫妮卡如何做美码和女码之间的换算。她说，美码裤子一般都偏瘦，通常相同大小的裤子，女码要比美码小3个尺寸。

比如，莫妮卡平时穿美码11码的裤子，换算成女码就是8码。当然，由于不同服装厂的裤子尺寸也会有差别，所以这个换算方法未必准确，但可以帮莫妮卡大概确定一个试穿尺码，省去一条条试穿的麻烦。

1. 请你帮珍妮计算一下，如果她通常穿美码17码的裤子，那她应该穿女码什么尺寸的裤子呢？

接下来，售货员又解释了美码和欧码尺寸之间的换算方法。以裤子为例，欧码裤子上的数字是指腰围的尺寸。比如，欧码26码的裤子，它的腰围就是26英寸。

售货员用卷尺量了莫妮卡的腰围是38.5英寸，然后给了她一个如下的换算图表：

美码	欧码
1	28
3	30
5	32
7	34
9	36
11	38
13	40
15	42

2. 那么莫妮卡应该先试一下什么尺寸的欧码裤子呢？

杰森也选中了几条裤子想试一下，男士尺码中美码、欧码是不同的，所以他不能用莫妮卡的方法来选尺码。

杰森选了一条欧码裤子，售货员告诉他欧码尺寸减掉16就是美码尺寸。

3. 杰森通常穿美码男士36码的裤子，而他手里选的这条裤子是欧码42码，那么杰森还需要试穿这条裤子吗？

杰森只需在他平时穿的美码男士尺码上加16，就能计算出他应该选择的欧码尺寸大小。

4. 那么，请你帮杰森计算一下，他应该试穿欧式尺寸什么码的裤子呢？

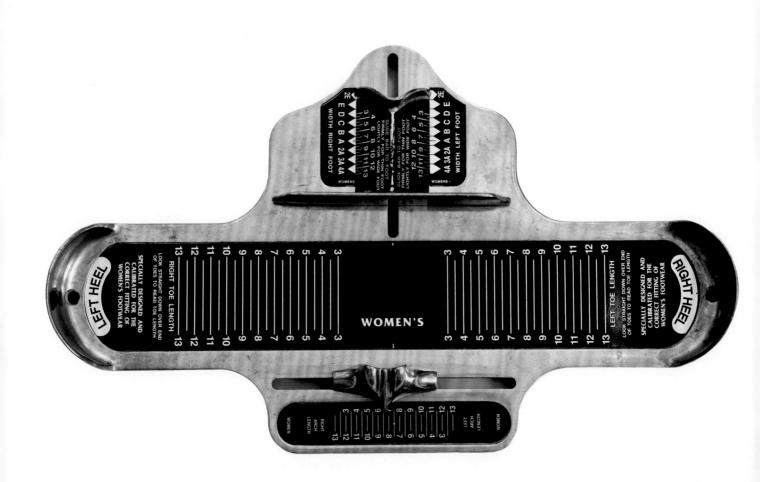

3
鞋子尺码与数学

逛完了服装店,三个好朋友又直奔鞋店。珍妮想买几双舞会上穿的鞋子,杰森也想看看有没有中意的运动鞋。

珍妮一直不确定自己应该穿的鞋子尺码,她以前买的鞋子要么大要么小,都不合适,总是弄得她的脚很痛。今天她是特意为舞会买鞋子,所以一定要选个合适的尺码。

鞋店有一个老式的脚长测量仪,莫妮卡以前也玩过,但从来没有用它量过自己脚的尺寸。她建议珍妮用这个测量仪量一下脚的大小,然后再问问售货员她穿什么尺码的比较合适。

下面是脚长和鞋子尺码对照表。

英寸	女码	男码
9	5	3 1/2
9 1/8	5 1/2	4
9 1/4	6	4 1/2
9 3/8	6 1/2	5
9 1/2	7	5 1/2
9 5/8	7 1/2	6
9 3/4	8	6 1/2
9 7/8	8 1/2	7
10	9	7 1/2
10 1/8	9 1/2	8
10 1/4	10	8 1/2
10 1/2	10 1/2	9

珍妮穿袜子测得的脚长是9 3/4英寸，虽然穿裙子时是不穿袜子的，但是她又不想光脚使用脚长测量仪，所以她还得做点小算术。

袜子大约增加了3/8英寸，所以珍妮还得在她的测量结果里减掉3/8英寸。

9 3/4和3/8分母不同，不能直接相减，我们还得将9 3/4转换成合适的分数形式：

$$9\,3/4 = 3\,9/4$$

然后将3 9/4转换成分母为8的分数。

1. 3 9/4 = _____ /8

现在用两个分数相减，然后再写成复分数。

2. 那么珍妮不穿袜子脚长是多少，你计算出来了吗？

3. 对比图表，珍妮应该穿多大码的鞋子呢？

4. 如果杰森脚长10 1/4英寸，那么他应该先试穿多大码的运动鞋呢？（当然，他穿运动鞋时是穿袜子的）

4
消费中的数学

莫妮卡和朋友们都有一些收获,莫妮卡买了一条裤子,珍妮买了一条裙子和几双鞋,杰森也还想再买几件T恤。

三个好朋友还想再逛一会儿,看看还有没有什么中意的东西,他们也想留点钱去吃冰激凌。

接下来,让我们算算如果他们还想留点钱吃冰激凌的话,那么还能再花多少钱?

莫妮卡最开始有56美元零花钱，然后买了一条裤子花了24.93美元，买了两副耳环，每副6.7美元，她还想再买一条项链，然后再留点钱吃冰激凌，她估计买冰激凌大概要花费6美元。

让我们计算一下莫妮卡一共花了多少钱，还剩下多少钱。

1. 如果莫妮卡还想留点钱去吃冰激凌，那么她在项链上最多能花多少钱？

杰森不记得一共花了多少钱，他的账单如下：

  太阳镜：14.56美元
  裤子：27.23美元

杰森计划一共消费不超过60美元，现在他不知道还能剩多少钱去买T恤。
如果每件T恤要7.5美元，那么他剩的钱能买几件T恤？
首先，你需要算算杰森还剩多少钱：

2.   60美元 - (14.56 + 27.23)美元 =

然后，你用剩余的钱数再除以每件T恤的价钱7.5美元，最后得到的整数部分就是答案。

3. 那么请你来算算，杰森用剩余的钱能买几件T恤呢？

4. 如果杰森还想留5美元买冰激凌，那么他还能买几件T恤呢？注意，他只能买整数件T恤哦（因为他不能买半件T恤呀！）。最终他还能剩多少钱呢？

5
促销中的数学

莫妮卡觉得自己花了太多钱,她不应该把零花钱全部花掉,应该留点钱日后再用。

莫妮卡还想买条价格实惠的项链。商场里到处都有9折、7.5折,甚至是5折的促销信息,她让朋友们帮她算一下折后价。

他们进了一家店,一眼就看到了打折标签。售货员告诉他们,不同颜色的标签代表不同的折扣:

蓝色:9.5折
橙色:8.5折
红色:7.5折
绿色:5.0折
黄色:2.5折

接下来让我们帮莫妮卡计算一下,哪个是她的最佳选择?

从上一节中，知道了莫妮卡用于买项链的资金额度是12.67美元，她想尽量不超过这个额度。她最初看中一条原价21.99美元的项链，折扣标签是红色的，那么莫妮卡能买得起么？

你可以用好几种方法来计算。红色标签代表折扣是7.5折。一种理解是，折扣后价格便宜了四分之一，用原价除以四，就得到了优惠的钱数。

1. 21.99美元/4=

这当然还没完，上面你得到了优惠的钱数，你还要再用原价减去这个数才能得到折后的价格。

2. 那么请你算算，这条项链的折后价是多少？莫妮卡能买得起么？

莫妮卡又看了另一条项链，原价55.60美元，虽然很贵，但是它挂着黄色的折扣标签，也就是2.5折！

这次，我们用交叉相乘法来算算她是否买得起这条项链。2.5折等同于优惠了75%的价钱。下面我们来计算一下：

$$75/100 = X/55.60 \text{美元}$$
$$100 \times X = 55.60 \text{美元} \times 75$$
$$X = 41.70 \text{美元}$$

项链是2.5折，也就是在原价的基础上，商家优惠了41.70美元。

你也可以将百分数写成小数，只需将小数点向左移两位，

$$75\% = 0.75$$

然后用原价乘以这个小数，得到的是优惠的价钱，原价再减去优惠的价钱就得到了折后价。

3. 0.75 × 55.60美元=

4. 那么请你来算算这条项链折后多少钱？莫妮卡买得起么？

5. 最后莫妮卡又看了另一条项链，原价是20.40美元，打5折。你可以用任意方法算算，莫妮卡能买得起这条项链么？如果买得起，那么她还能剩下多少钱？

6
在线购物中的数学

逛完街回到家,莫妮卡突然想起忘了给爸爸买生日礼物。爸爸告诉她,自己的领带都旧了,想要一条新领带。莫妮卡不想再去一趟商场了,所以她准备在网上给爸爸选一条领带。

莫妮卡选中了一条不错的领带,准备买下来。这条领带要16.99美元,一开始莫妮卡还觉得价钱可以,后来她意识到还要交消费税和快递费。消费税是国家向消费者征收的,快递费是把物品送到买家而收的运输费用。

莫妮卡没有自己的储蓄卡和信用卡,所以她请妈妈帮忙在网上付款。莫妮卡逛街差不多把零花钱都花光了,所以她还得跟妈妈借钱买这份礼物。接下来,我们来算算莫妮卡需要借多少钱,她又是怎么还钱的。

莫妮卡所在州的消费税是9.5%。百分数的算法跟前一节讲过的一样，用原价乘以百分数得到一个数，前一节讲过是用原价减掉这个数，在这一节，我们是用原价加上这个数。

1. 那么请你来算算，算上消费税，买这条领带要花多少钱呢？

接下来，你还要再加上快递费，然后才能得到在网上要付的总价是多少？

2. 如果快递费是4.99美元，那么买这条领带要付的总价是多少？

莫妮卡现在手里没有足够多的钱还给妈妈。

3. 在第5节里，我们知道莫妮卡还剩了一点钱，她把剩下的钱都还给妈妈后，还欠多少钱？

莫妮卡同意以分期还款的形式还妈妈钱。她每周还给妈妈5美元，这样她就有足够的时间来攒钱。

看下表，请你将表格补充完整，并算出莫妮卡多少周后才能还清欠款。

周数	欠款额/美元	待还款额/美元
1	23.59	18.59
2	18.59	13.59
3	13.59	

4. 请你算算莫妮卡要用多长时间才能还清欠款？

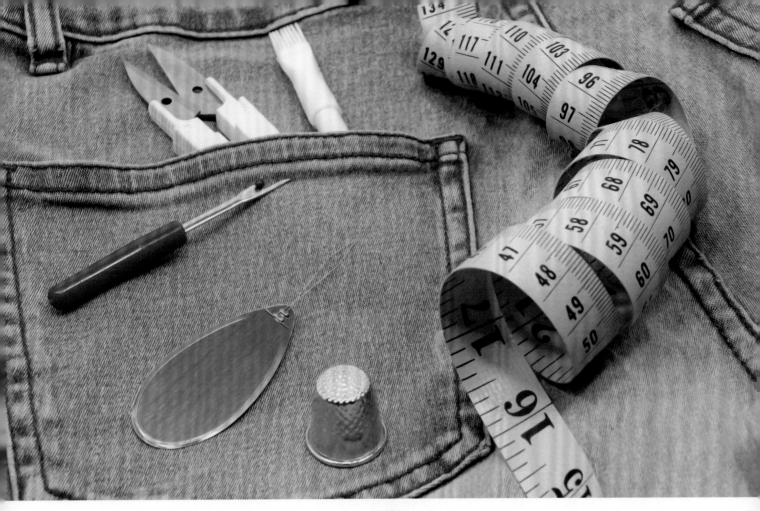

7
量体裁衣中的数学

莫妮卡在家又试了一下新买的裤子,她发现裤腰肥瘦刚刚好,只是裤腿太长了。她试着把裤脚长的部分折过来,但是看起来又挺傻的。

莫妮卡已经把裤子的标签撕掉了,所以她也不能把裤子退掉了。妈妈建议她去找街角的裁缝帮忙处理一下。妈妈说裁缝可以把裤腿剪掉一截,这样裤腿长度就合适了,这个做法专业术语叫钎边。

莫妮卡和妈妈拿上裤子去了裁缝铺。妈妈的夹克衫有点儿瘦了,又不想扔掉,所以也带上夹克衫请裁缝看看能不能改松一点儿。

裁缝孙师傅把莫妮卡和妈妈带到后屋,先给莫妮卡量了一下腿长,然后又给妈妈量了一下腰围,也就是夹克衫有些紧的地方,她说裤子和夹克衫很容易改,明天来取就行了。

首先，先处理裤子，孙师傅量了莫妮卡从裤裆到脚踝的长度是28 1/2英寸。

莫妮卡和妈妈离开后，孙师傅量了莫妮卡裤子从裤裆到裤腿的长度，是30 3/4英寸。

1. 请你算算莫妮卡实际腿长和裤腿长度相差多少？

上面得到的结果就是孙师傅要将裤腿钎起来的长度。她想把裤脚折起来，这样裤脚就能改到合适的长度了。

孙师傅将裤脚折叠两次，这样原来的裤边就被隐藏起来了。

孙师傅每折叠一次要卷多宽的裤脚？你要将问题1得到的答案写成分数，然后再除以2（因为裤脚是折叠了两次），最后再写成合适的分数形式，这样就得到了每次折叠的宽度。

2. 那么你来算算吧？

孙师傅将折叠好的裤脚缝好，这样莫妮卡的裤子钎边完成了！

接下来，孙师傅要把莫妮卡妈妈夹克衫腰围从接缝处放出来一些，这样就能宽松一些了。

莫妮卡妈妈的腰围大约是41英寸，夹克衫扣上扣子后腰围大约是41 1/4英寸，孙师傅想将腰围放宽3/4英寸，这样衣服就不会这么紧了。

实际做起来并不像听起来那么难。孙师傅只需将腰围处的接缝拆开，然后在更靠近边缘的地方缝上，两边都这样做，夹克衫就会宽松一点儿了。

3. 如果孙师傅想一共放出3/4英寸，并且衣服两边接缝处放出来的宽度应该是一样的，那么每边应该放出多宽呢？

8
处理旧衣服中的数学

当裁缝在给莫妮卡裤子钎边的时候,莫妮卡整理了一下自己的衣柜。她发现自己有太多衣服了,甚至有一些衣服都没怎么穿过。衣柜最里面有几件衬衫看起来很不错,为什么之前没有穿过呢,莫妮卡自己也很纳闷。她把这些衣服翻出来试了一下,结果这些衣服都是几年前买的了,现在已经不合身,太小了。

莫妮卡开始将那些小了的或不会再穿的裤子、衬衫等衣服打包然后处理掉。

莫妮卡并不想把这些旧衣服扔掉,因为这些衣服还可以穿。然后她想起来表哥曾告诉她可以把这些旧衣服送到寄卖店,寄卖店会将这些旧衣服再次出售。寄卖店会将表哥那些看起来还不错的旧衣服留下,一旦这些旧衣服卖出去了,他就能得到一笔钱。

莫妮卡又想到了自己经常去的二手衣店。她可以把旧衣服送到二手衣店,他们会买下那些看起来还不错的旧衣服。虽然二手衣店挑选旧衣服比较苛刻,但莫妮卡还是有几件很不错的衣服。接下来我们看看莫妮卡这些旧衣服可以卖多少钱。

莫妮卡先去了二手衣店，她看了一下衣服价格：

T恤：3美元
衬衫：4美元
毛衫：5美元
裤子：5美元
半身裙：4美元
连衣裙：6美元
首饰：3美元
夹克衫：10美元

莫妮卡带去了3条裤子、4件T恤、1件衬衫、2条连衣裙和1条项链。

1. 请你来算算，在二手衣店里，莫妮卡最多能卖出多少钱？

售货员仔细查看了莫妮卡带去的旧衣服，最后收下了1条裤子、3件T恤、1条连衣裙和那条项链。

2. 那么，莫妮卡赚到多少钱呢？

接下来，莫妮卡要去寄卖店。寄卖店也收一些看起来不是很好的衣服，所以莫妮卡打算把剩下的旧衣服都送到寄卖店。
寄卖店的售货员说每卖出一件衣服，莫妮卡可以得到售价的20%。
寄卖店收下了莫妮卡的旧衣服，售货员给每件衣服定售价：

2条裤子：5.99美元/条
1件T恤：2.99美元
1件衬衫：4.50美元
1条连衣裙：8.75美元

3. 那么请你帮莫妮卡算算，如果她的这些旧衣服都能卖出去，那么她能赚到多少钱呢？

9
裁剪圆形中的数学

自从莫妮卡去过裁缝铺后,她就想,要是自己会做衣服该有多棒啊!她很喜欢艺术,做过一些针织和钩编。莫妮卡想如果自己会做衣服的话,就可以省下一笔买衣服的钱,还可以做一些特别的设计。

莫妮卡自学了裁剪,然后决定自己试着做一条半身短裙。她只需先去选一块面料,然后稍稍剪裁一下。她想的简单做法就是剪出一个圆形,然后在中心再减掉一个和自己腰围同样大的圆,最后再缝上一条腰带,做上钎边,半身短裙就做好了。

教材书上面说先要确定半身裙的长度,莫妮卡决定做一条18英寸长的裙子。现在她要做一些小小的算术,看看做这样的一条裙子需要多少面料。莫妮卡的办法是先算出圆形的半径,然后就能得到圆形的面积,这样就她就知道要买多少面料了。

要做一条长度是18英寸的半身短裙，莫妮卡需要以小圆的圆边为起点，往外量出18英寸的长度，然后再做出一个大圆。

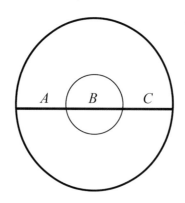

首先莫妮卡要先量量自己的腰围。在商场试衣服时，她知道了自己的腰围大约是38.5英寸。接下来她要算算周长是38.5英寸的圆，直径是多少。

我们知道，直径和周长之间的公式是：

$$C = \pi \times d$$

C代表圆的周长，d代表圆的直径，π是圆周率3.14。接下来你就转换这个公式算出直径了：

$$d = C/\pi$$

1. 请你来算算莫妮卡的腰围直径是多少？

现在莫妮卡知道了自己要在面料中间减掉多大的圆了（如圆B），接下来，她还得算出外面的大圆要多大。

莫妮卡想要短裙的长度是18英寸，那么图中A和C的长度至少应为18英寸，由于还要在边缘处留出做钎边的长度，所以每边至少还要增加0.5英寸。

现在，莫妮卡要把所有数据整理起来，算出大圆的直径是多少。

2. 请你帮莫妮卡算算，大圆的直径是多少呢？

10
裁剪矩形中的数学

莫妮卡对剪裁非常感兴趣。除了简单的半身短裙外,她现在想去尝试做一个简单的包,她的朋友约翰就自己做了一个背包。莫妮卡觉得背包对她来说还是有点难,所以她想自己可以试试做一个小钱包。

莫妮卡在网上看到了几种图样,她觉得可以把几块长方形缝起来做成一个包。她想自己做设计,以便看看自己是不是会做了。她拿出了一张纸,试着在纸上裁出几块长方形。

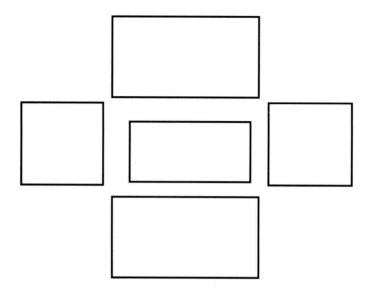

莫妮卡设计的样式很简单，但还是要做一些算术。她需要计算出这几块长方形的大小，这样才能将它们缝起来。

她先计算包底的部分，也就是中间的长方形。她计划做一个小包，所以她将包底的长方形设计为长8英寸、宽4英寸。

接下来，莫妮卡突然想到她还要将两块长方形缝上，那么接缝处还要占一定的长度，所以莫妮卡每边又增加了1/4英寸，最终包底的长方形长为8 1/4英寸，宽为4 1/4英寸。

其他的长方形要相应地缝在中间长方形四周。莫妮卡计划包的高度是5英寸，这样也就决定了其他长方形的大小。

由于包的高度是5英寸，所以两边小长方形的长应为5 1/4英寸；另一边要与底部长方形相接，所以宽应为4 1/4英寸。

1. 那么请你来算算，上下两个大长方形的长应为多少呢？

2. 它们的宽应为多少呢？

莫妮卡还得再做一个长24英寸、宽2 1/2英寸的包带。

11
购买面料中的数学

莫妮卡在动手做短裙和钱包之前,她还得做些算术,要算出买多少面料。如果面料买多了,那就太浪费钱了。如果买少了,又做不成短裙和钱包。

莫妮卡要通过尺寸和面积来计算需要面料的大小。通过直径可以计算出需要做短裙面料的大小,通过长方形的长宽可以计算出需要做钱包面料的大小。那么我们来看看莫妮卡是如何计算的呢?

莫妮卡在做短裙之前要先剪出一个圆，通过第9节你可以知道那个圆的直径是多少。

现在我们来算算需要多少面料吧！扎成一捆的布料宽是45英寸，那么莫妮卡要买多长呢？

1. 请你先算算布料的宽45英寸是否比圆的直径大？如果足够大，那么多出多少？

一块宽45英寸、长1码的布料是否足够做短裙呢？一个方法就是看看1码是否比圆的直径长，可以用如下等式将1码转换成英寸：

$$1码=3英尺, \quad 1英尺=12英寸$$

2. 1码是否足够长呢？如果不够，那么莫妮卡还需要多长布料呢？

当然也有四分之一码和半码，四分之一码就是：

$$0.25码 \times 3 \times 12 = 9英寸$$

3. $1\frac{1}{4}$码与圆的直径比较，是否够大呢？

现在还要买一块布料做钱包，她知道长方形面积的计算公式是：

$$A = 长 \times 宽$$

莫妮卡需要将要剪裁的所有长方形的面积加起来，然后买一块至少面积要大于所有长方形面积和的布料。当然你知道的，1/4英寸就是0.25英寸。

4. 那么请你来算算，所有长方形的面积和是多少平方英寸呢？

现在请你计算一下，宽45英寸、长半码的布料面积是多少？

5. 45英寸 × (0.5码，× 3英尺，× 12英寸)=

6. 这块布料是否够做钱包呢？如果不够，那还需要多少？如果再额外多1/4码，是否足够呢？

12
服装成本中的数学

莫妮卡越来越喜欢自己做衣服了，而且她也发现自己很擅长做衣服。她的朋友们也都很喜欢她做的衣服，慢慢地朋友们都来请莫妮卡帮忙做衣服了。

最初莫妮卡真的很喜欢帮朋友们做衣服。一方面她能练习剪裁技术，另一方面又能使朋友们很开心。然而，慢慢地莫妮卡发现要去买原材料真的很耽误时间，而且如果原材料还要她自己出钱购买的话，那莫妮卡很快就要破产啦！

所以，莫妮卡计划通过卖自己做的衣服赚一些钱。她可以在手工集市上卖衣服，也可建个网站，在网络上卖。在开始销售之前，莫妮卡还得算一下她做的衣服成本价是多少钱，她要将所有原材料的价钱加起来，为了获得利润，售价当然要高于成本价。另外，莫妮卡还要想想做衣服花费的时间和技术的成本是多少，这些也都要算到售价中。

莫妮卡将原材料列了一个清单，这样来计算做衣服的成本。

下面是莫妮卡最近做一条连衣裙的材料清单

材料	费用/美元
纽扣	4.98
线	7.45
布料	26.23

1. 那么做一条连衣裙的成本是多少呢？

莫妮卡又从另一角度计算了一下成本，她不需要为做连衣裙特别地去买针线和纽扣，她可以使用现有的针线和纽扣。因此，连衣裙的成本应该只有布料钱。

2. 你认为莫妮卡卖一条连衣裙至少要价多少钱才能保证收益？

现在莫妮卡要计算一下她的时间成本，然后才能确定衣服的总价是多少。

莫妮卡做一条连衣裙要花整整15个小时，这个时间她本可以用来打工赚钱。事实上，她的朋友们有的在商场打工，有的在餐馆打工，他们大概每小时能赚9美元。

3. 请你来算算，在商场打工每小时可以赚9美元，如果莫妮卡用做衣服的时间去商场打工的话，那么她可以赚到多少钱？

莫妮卡可以在布料成本基础上增加多少钱呢，她想看看这样的价钱是否合理。

4. 你认为顾客愿意花这个价钱买这条连衣裙么？是太便宜了还是太贵了呢？

莫妮卡还想让连衣裙的价钱更便宜一些，这样才能吸引更多的顾客。她还是一个新手裁缝，以后随着技术越来越成熟，卖衣服的价钱可以更高一些。

莫妮卡决定将连衣裙的售价定为总价的35%。

5. 请你来算算，连衣裙的最新售价是多少？你觉得这个价钱对顾客和莫妮卡来说是否都是合理的呢？

13
销售提成中的数学

莫妮卡刚刚听说镇里新开了一家商店，可以寄卖人们自己做的手工艺品，如画作、剪纸、衣服等。在这家商店销售自己的手工作品，商店要收一些手续费，也就是销售提成，剩下的才归制作人。莫妮卡从未想过她做的衣服还能在商店里销售。

莫妮卡去找商店经理汤姆谈代售的事情。汤姆告诉她，商店会从她的售价中抽成15%作为销售提成，换句话说，每卖出一件商品，商品将收取售价的15%。而且商店会给莫妮卡的衣服定价，因为他们更了解顾客的消费意向。

商店的定价会不会太低了呢，莫妮卡对此有点儿担忧。其实她的担心是多余的，因为汤姆告诉她商店给莫妮卡衣服的定价是连衣裙75美元/件，衬衫50美元/件，半身短裙50美元/件，包40美元/件。这个定价都高于莫妮卡之前的卖价，莫妮卡真是一个不错的设计师和好裁缝。

请你填写下面表格，并计算一下如果莫妮卡的衣服都卖出去了，她能赚多少钱？

物品	售价/美元	商店抽成/美元	剩余/美元
连衣裙	75	11.25	63.75
衬衫			
半身短裙			
包			

1. 与莫妮卡自己卖衣服相比，在商店里卖赚的钱是多了么？如果是多了，那多多少？如果是少了，少多少呢？

2. 如果莫妮卡做一个包要花16.50美元买布料，那么她能赚多少钱呢？

3. 莫妮卡卖出多少件衬衫才能赚到至少150美元呢？

14
在线销售中的数学

在商店里卖衣服很有意思,但莫妮卡还想接触到更多的人。只有去商店里的人才能看到她的衣服,但莫妮卡想会有更多的人有兴趣购买自己做的衣服。莫妮卡准备在网上开店卖衣服,这也正是个好机会!

首先,莫妮卡要创建一个自己的网站,而且最初莫妮卡要非常努力才能让人们先去浏览她的店铺。另一个做法是,莫妮卡可以在网上的虚拟商场上卖衣服,虚拟商场就是为艺术家和手工制作人卖东西而创建的一个公共网站。

莫妮卡在网站上进行了注册并阅读了网站规则。莫妮卡在网站上每挂上一件物品要付给网站0.2美元,每卖出一件物品,网站将要抽成3.5%作为销售提成。莫妮卡不得不去计算将物品送到买家手里的快递费。这么多费用,接下来我们看看莫妮卡是怎么计算的吧!

首先，莫妮卡要给出网上物品的售价，由于在网上卖物品可以自己定价，所以莫妮卡决定就以实体商店给的售价为准。

以75美元/件的连衣裙为例，去掉网站的抽成，那么莫妮卡能得到多少钱？

成本：
连衣裙挂在网站上要收取0.2美元的费用
3.5% × 75美元 = 2.63美元

1. 莫妮卡卖出一件连衣裙能最终收到多少钱呢？如果购买连衣裙布料要花28.30美元，那么莫妮卡的净收益是多少呢？

现在我们用以下信息来计算一下快递费：

- 一件连衣裙重15盎司
- 包装盒重6盎司
- 包装盒售价1.99美元
- 邮局对每盎司收费0.35美元

那么莫妮卡邮寄一件连衣裙要花费多少钱呢？

总重量：15盎司+6盎司=21盎司
邮寄费用：21盎司 × 0.35美元/盎司=7.35美元
加上包装盒后的总邮费：7.35美元+1.99美元=9.34美元

2. 算上邮寄费，这位顾客要付多少钱呢？

如果有海外顾客要购买这条连衣裙，那么邮寄费用的底价是12美元，每一盎司0.6美元，显然邮寄费用要贵很多。

3. 那么海外顾客购买这条连衣裙要花多少钱呢？

15
小　结

和过去只知道去商场购物相比，近来莫妮卡学到了很多知识，包括如何剪裁衣服、如何计算成本、如何设计衣服、如何把衣服卖给顾客。

未来，莫妮卡计划去学校专门学习设计，然后在时尚界闯出名气。现在她已经掌握了这么多数学知识，充满了自信。让我们来看看你是否也掌握了这些知识呢。

1. 如果一个女孩平时穿欧码32号的牛仔裤，那么美码应该穿多大号的？

2. 如果一个男孩穿袜子后脚长9 3/4英寸，那么他应该穿多大码的鞋呢？

3. 假设你想买一件运动衫，商场里有一件运动衫打6折，原价是32.99美元，请你算算打折后是多少钱？

4. 如果买这件运动衫，还需要另付9.5%的消费税，如果你手里只有25美元，能够买下这件运动衫吗？为什么呢？

5. 如果一条裤子裤腿长是29英寸，如果某人从裤裆到裤脚的长度是26英寸，那么这人穿这条裤子是否需要钎边？

如果需要钎边，那么钎多少呢？

6. 如果你有一件旧衬衫要拿去寄卖，定价是6.5美元，售出后会分给你定价的20%。那么这件衬衫卖出后你能拿到多少钱？

7. 你想裁制一件半身裙，需要825平方英寸的布料，商场里的布料宽是45英寸，但只有半码长了，请你计算一下，这块布料够用吗？

如果不够用，还需要多少布料？

8. 如果你花88美元买块布料裁成衣物，然后卖出20美元。

那么，你的收益是多少？

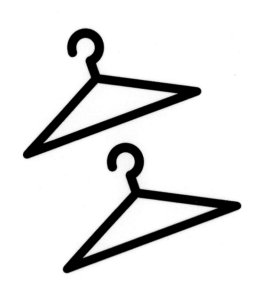

参考答案

1.
1. (9 + 11 + 11 + 13 + 15)/5 = 11.8
2. 11
3. 平均值是2，他穿M码
4. M码

2.
1. 14码
2. 38
3. 不需要，因为换算为美码是男士26码
4. 欧式52码

3.
1. 78
2. 78/8 - 3/8 = 75/8
   75/8 = 9 3/8 英寸
3. 9 1/2
4. 8 1/2

4.
1. 56 - (24.93 + 6.70 + 6.70 + 6) = 11.67
2. 18.21美元
3. 2
4. 一件，如果他买了1件T恤，则还剩10.71美元

5.
1. 5.50美元
2. 16.49美元。她没有足够的钱，买不起
3. 41.70美元

4. 13.90美元。她买不起
5. 是的,她买得起,因为折后只需10.20美元,买完项链后还剩2.47美元

6.

1. 16.99 × 0.095 = 1.61, 16.99 + 1.61 = 18.60
2. 18.60 + 4.99 = 23.59
3. 23.59 - 2.47 = 21.12
4. 5周

周数	欠款额/美元	待还款额/美元
1	23.59	18.59
2	18.59	13.59
3	13.59	8.59
4	8.59	3.59
5	3.59	0

7.

1. $2\frac{1}{4}$ 英寸
2. 每次折叠 $1\frac{1}{8}$ 英寸 ($2\frac{1}{4} = 9/4$, $9/4 / 2 = 9/8 = 1\frac{1}{8}$)
3. 3/4 英寸 / 2 = 3/8 英寸

8.

1. 46
2. 23
3. 0.20 × (2 × 5.99 + 2.99 + 4.50 + 8.75) = 5.64

9.

1. d = 38.5 英寸 / 3.14
   d = 12.26 英寸
2. 0.5英寸 + 18英寸 + 12.26英寸 + 18英寸 + 0.5英寸 = 49.26英寸

10.

1. 为了匹配中间长方形的长度，上下两个大长方形的长应是 8 1/4 英寸
2. 为了匹配包的长度，宽应为 5 1/4 英寸

11.

1. 不，这块布料不够大
2. 不，她还需要13.26英寸长的布料(1码 × 3英尺 × 12英寸 = 36英寸)
3. 不
4. (2 × (5.25 × 4.25)) + (2 × (8.25 × 5.25) + (8.25 × 4.25) + (24 × 2.5)) = 226.31平方英寸
5. 810平方英寸
6. 这块布料足够了，如果再额外多1/4码，那就是又多了405平方英寸了

12.

1. 38.66美元
2. 她至少要卖出26.24美元才能获利
3. 135美元
4. 161.23美元。连衣裙有点儿贵，也许会有人来买
5. 56.43美元。是的，这个价位看起来更合理一些

13.

1. 放在商店里卖的多赚7.32美元
2. 17.50美元
3. 4件衬衫

物品	售价/美元	商店抽成/美元	剩余/美元
连衣裙	75	11.25	63.75
衬衫	50	7.50	42.50
半身短裙	50	7.50	42.50
包	40	6	34

14.

1. 72.17美元, 43.87美元
2. 84.34美元
3. 24.60美元

15.

1. 美码5码.
2. 6 1/2
3. 19.80美元 (0.4 × 32.99 = 13.19美元, 32.99 - 13.19 = 19.80)
4. 是的，你有足够的钱买这件运动服，还剩余3块多美元(19.80 × 0.095=1.88, 19.80 + 1.88 = 21.68美元)
5. 是的, 需要纤3英寸
6. 6.50 × 0.2 = 1.30美元
7. 不够用，还需再长0.3英寸
   布料长度：
   0.5码 × 3英尺 × 12英寸 = 18英寸
   需要长度：
   825平方英寸 = 长 × 45英寸
   长 = 18.3英寸
8. 11.20美元

INTRODUCTION

How would you define math? It's not as easy as you might think. We know math has to do with numbers. We often think of it as a part, if not the basis, for the sciences, especially natural science, engineering, and medicine. When we think of math, most of us imagine equations and blackboards, formulas and textbooks.

But math is actually far bigger than that. Think about examples like Polykleitos, the fifth-century Greek sculptor, who used math to sculpt the "perfect" male nude. Or remember Leonardo da Vinci? He used geometry—what he called "golden rectangles," rectangles whose dimensions were visually pleasing—to create his famous *Mona Lisa*.

Math and art? Yes, exactly! Mathematics is essential to disciplines as diverse as medicine and the fine arts. Counting, calculation, measurement, and the study of shapes and the motions of physical objects: all these are woven into music and games, science and architecture. In fact, math developed out of everyday necessity, as a way to talk about the world around us. Math gives us a way to perceive the real world—and then allows us to manipulate the world in practical ways.

For example, as soon as two people come together to build something, they need a language to talk about the materials they'll be working with and the object that they would like to build. Imagine trying to build something—anything—without a ruler, without any way of telling someone else a measurement, or even without being able to communicate what the thing will look like when it's done!

The truth is: We use math every day, even when we don't realize that we are. We use it when we go shopping, when we play sports, when we look at the clock, when we travel, when we run a business, and even when we cook. Whether we realize it or not, we use it in countless other ordinary activities as well. Math is pretty much a 24/7 activity!

And yet lots of us think we hate math. We imagine math as the practice of dusty, old college professors writing out calculations endlessly. We have this idea in our heads that math has nothing to do with real life, and we tell ourselves that it's something we don't need to worry about outside of math class, out there in the real world.

But here's the reality: Math helps us do better in many areas of life. Adults who don't understand basic math applications run into lots of problems. The Federal Reserve, for example, found that people who went bankrupt had an average of one and a half times more debt than their income—in other words, if they were making $24,000 per year, they had an average debt of $36,000. There's a basic subtraction problem there that should have told them they were in trouble long before they had to file for bankruptcy!

As an adult, your career—whatever it is—will depend in part on your ability to calculate mathematically. Without math skills, you won't be able to become a scientist or a nurse, an engineer or a computer specialist. You won't be able to get a business degree—or work as a waitress, a construction worker, or at a checkout counter.

Every kind of sport requires math too. From scoring to strategy, you need to understand math—so whether you want to watch a football game on television or become a first-class athlete yourself, math skills will improve your experience.

And then there's the world of computers. All businesses today—from farmers to factories, from restaurants to hair salons—have at least one computer. Gigabytes, data, spreadsheets, and programming all require math comprehension. Sure, there are a lot of automated math functions you can use on your computer, but you need to be able to understand how to use them, and you need to be able to understand the results.

This kind of math is a skill we realize we need only when we are in a situation where we are required to do a quick calculation. Then we sometimes end up scratching our heads, not quite sure how to apply the math we learned in school to the real-life scenario. The books in this series will give you practice applying math to real-life situations, so that you can be ahead of the game. They'll get you started—but to learn more, you'll have to pay attention in math class and do your homework. There's no way around that.

But for the rest of your life—pretty much 24/7—you'll be glad you did!

1
SIZE MATH

Maricela and a bunch of her friends are on a shopping trip. They all decide to go to the mall together, for a fun day. Maricela needs a few clothes too—she needs a new pair of jeans. Her friend Jason is looking for some t-shirts. And her friend Jacqui has to buy a new dress for a school dance.

Maricela really likes going shopping, because she gets to hang out with her friends. However, she gets frustrated sometimes when she can't find anything that fits. She spends hours in one store trying on dozens of clothes, since she never knows what will fit.

One way she could think about clothes sizes is in averages. If she knows her average clothing size, she can more easily find clothes that might fit her. On the next pages, you can explore averages and clothing size.

Maricela thinks of the last few times she's bought jeans. She has jeans of different sizes at home. She has one pair that is a junior's size 9, two that are size 11, one that is size 13, and one that is size 15.

To find the average size of her jeans, add up all the sizes and divide by the number of jeans.

1. What is Maricela's average jean size?

The average you get when you do the math is a decimal, not a whole number. Clothing sizes only come in whole numbers. In Maricela's case, she wears junior's sizes, which only come in odd numbers. In this case, her average size will be the odd number closest to her calculated average.

2. What is the closest junior's size to Maricela's calculated average?

Now Maricela knows which size to start trying on first.

You can also find the average sizes for the clothes her friends are trying to buy.

Jason has 3 t-shirts at home in small, 5 in medium, 1 in large, and 1 in extra large. You can't do averages using words, but you can assign each size a number.

$$\text{Small} = 1$$
$$\text{Medium} = 2$$
$$\text{Large} = 3$$
$$\text{Extra large} = 4$$

Now do the average.

3. What is Jason's average t-shirt size? You may have to round to the nearest whole number.

Now find the average dress size Jacqui should be looking for. She has 1 dress at home that is small, 1 that is medium, and 3 that are large.

4. What is Jacqui's average dress size?

2
MORE SIZE MATH: CONVERSIONS

In the second store Maricela and her friends go into, she finds three pairs of jeans she wants to try on. The trouble is, she doesn't recognize the sizes. She normally wears a size 11 in juniors, or sometimes a size 13. These jeans don't come in junior's sizes. Instead, two of the pairs she wants to try on are in women's sizes, and one is in European sizing.

Maricela is worried she'll have to try on every pair of jeans in the store to find one that fits. However, a store clerk comes over and offers to help. The store clerk tells her how to **convert** junior's sizes to women's and European sizes. Women's sizes are slightly smaller, and European sizes are based on waist measurement. Check out the next page to follow what Maricela learned.

First, the store clerk shows Maricela how to convert junior's sizes to women's sizes. She explains that junior's jeans are usually slimmer. She tells Maricela that women's sizes are usually about 3 numbers smaller than junior's sizes.

So, if Maricela usually wears an 11, she may wear an 8 in women's sizes. Of course, that won't always be true because the size of jeans is a little different from company to company. But it gives Maricela a place to start.

1. What size in women's jeans should Jacqui wear, if she normally wears a size 17 in juniors?

Now the store clerk explains how European sizing works. For jeans, the size number is how big around the waist is. Someone with a 26-inch waist would wear a size 26.

The store clerk takes out a tape measure and measures Maricela's waist, which is 38.5 inches. Then she shows her a chart that looks like this:

39

U.S. Junior's size	European size
1	28
3	30
5	32
7	34
9	36
11	38
13	40
15	42

2. What European size in jeans should Maricela try on first?

Jason decides he wants to try on some jeans too. Men's sizes are different from women's and juniors, though, so he can't follow the same rules as Maricela.

He finds a pair of jeans in a European size. The sales clerk tells him he can subtract 16 from the size on the tag to find the U.S. men's size.

3. Jason normally wears a size 36 in men's. The tag says the jeans he's holding are a size 42. Should he try them on? Why or why not?

Jason can add 16 to his usual size to figure out what European size he should look for.

4. What size should he look for?

3
SHOE SIZE

After the clothing store, the three friends head over to a shoe store. Jacqui needs some shoes for the dance, and Jason wouldn't mind some new sneakers if he can find them. Jacqui can never figure out what shoe size she should wear. She always seems to buy the wrong size, and then her feet hurt. She's going to be dancing in the shoes she buys today, so she wants to make sure she gets the right size.

The shoe store has an old-fashioned foot measure. Maricela has played around with one before, but never really used it to measure her foot. She hands it over to Jacqui and suggests she uses it to measure how big her foot is. Then they can ask the sales clerk what size shoe she should get.

Here's a chart that shows shoe size and how big each size is in inches.

Inches	Girl's/women's sizes	Boy's/men's sizes
9	5	3½
9 ⅛	5½	4
9¼	6	4½
9 ⅜	6½	5
9½	7	5½
9 ⅝	7½	6
9¾	8	6½
9 ⅞	8½	7
10	9	7½
10 ⅛	9½	8
10¼	10	8½
10½	10½	9

Jacqui measures out her foot. It is 9¾ inches. However, she measured her foot in socks, and she won't be wearing socks with her dress shoes. She doesn't really want to put her bare foot on the measuring tool, so she uses math instead.

Socks add about 3/8 of an inch to your foot, so Jacqui has to subtract 3/8 of an inch from her foot measurement.

9¾ inches and 3/8 inch do not have common denominators. You'll have to make the first number into an improper fraction:

9¾ = 39/4

The common denominator between the numbers you're adding is 8.

1. $^{39}/_4 =$ _____ $/8$

   Now use the subtraction and change it back into a mixed fraction.

2. How big is Jacqui's foot without her sock?

3. According to the chart, what size shoe should Jacqui wear?

4. If Jason's foot is 10¼ inches long, what size sneaker should he try on? (He'll be wearing socks with his sneakers.)

4
SPENDING MONEY

Maricela and her friends have found a few things to buy, and they have spent some of their money. Maricela found a pair of jeans. Jacqui bought a dress and some shoes to go with it. Jason hasn't had as much luck and still needs a few new t-shirts.

The three friends aren't ready to leave the mall yet. They want to keep shopping for a little while. They have more to spend, although they want to save a little at the end to get ice cream in the food court.

Figure out how much they can spend on the next page, and if they will have enough left over for ice cream.

Maricela started out with $56 she had saved up from her allowance. She spent $24.93 on a pair of jeans, and also bought two pairs of earrings for $6.70 each. She still wants to buy a scarf, and have money left over for ice cream, which she thinks will be about $6.

Add up all the money she has spent, and subtract it from the total money she has to spend.

1. How much can Maricela spend on the scarf if she wants money left over for ice cream?

Jason can't quite remember how much money he has spent so far. He takes a look at his receipts and adds them up.

He finds receipts for:

> sunglasses: $14.56
> jeans: $27.23

He wants to spend less than $60, but he's not sure how much he has left to spend on t-shirts.
How many t-shirts can he buy with the money he has left, if they cost $7.50 each?
First, you need to see how much money he has left:

2. $60 − ($14.56 + $27.23) =

Next, divide your answer by $7.50, the cost of a t-shirt. Round down to the nearest whole number.

3. How many t-shirts can he buy with the money he has left?

4. How many t-shirts can he buy if he wants to save at least $5 for ice cream? Remember, he can only buy t-shirts in whole numbers (since he can't buy half a t-shirt!). How much will he have left in total?

5
SALES MATH

Maricela is starting to worry she's spending too much money. Maybe she shouldn't spend everything she has saved up for the shopping trip. She might need it later for something else.

She still wants to buy a scarf, though. She thinks her best bet is to find a good sale. Everywhere she's gone in the mall, she has seen signs for sales of 10%, 25%, and even 50% off. She asks her friends to help her on her sale hunt.

The next store they go into, they see sale signs right away. The signs tell them that clothes with different colored tags have different sales prices:

Blue: 5% off

43

Orange: 15% off
Red: 25% off
Green: 50% off
Yellow: 75% off

What is the best deal? Find out on the next page.

You found in the last section that Maricela has $12.67 to spend on a scarf. She wants to spend less than that if possible. Maricela first finds a scarf that originally cost $21.99. It has a red tag. Can she afford it?

You can think about percents in a few ways. The red tag means the scarf is 25% off. Another way of think about 25% is one-fourth of something. The price of the scarf costs one-fourth less. Divide the price by 4 to find out the discount.

1. $21.99/4 =

You're not done yet. Now you know how many dollars less the scarf is. You still have to subtract the number you go from the full price of the scarf.

2. What is the final discounted price? Can Maricela buy it?

Maricela finds another scarf that was $55.60, but has a yellow tag. It used to be expensive, but now it's 75% off!

This time, use cross multiplication to figure out if she can buy it: 75% off is the same as saying 75 out of 100. Here's how you figure it out:

$$75/100 = X/\$55.60$$
$$100 \times X = \$55.60 \times 75$$
$$X = \$41.70$$

The scarf is 75% off, so you have to subtract $41.70 from the original price.

You could also convert the percent to a decimal. Just move the decimal place to the left two spaces.

$$75\% = 0.75$$

Then multiply the cost by the percent, and subtract that number from the original cost. Try it:

3. 0.75 x $55.60 =

4. How much money does this scarf cost? Can she afford it?

5. Finally, Maricela finds another scarf that cost $20.40, and is 50% off. Can she buy this one? How much will she have left if she can buy it? Use whatever method you want to figure it out.

6 SHOPPING ONLINE

When Maricela gets home from shopping at the mall, she realizes she forgot to buy her dad a birthday present! He told her he needs a new tie, because all his other ones are wearing out. She really doesn't want to go back to the mall. Instead, she sits down at the computer to find her dad a tie online.

Once Maricela finds the perfect tie, she's ready to buy it. The tie Maricela picks costs $16.99 online. She thinks she's getting a good deal until she remembers she has to pay tax and shipping and handling. Tax is money the government charges and collects every time someone makes a purchase. Shipping and handling is the money she'll have to pay to get the tie sent to her in the mail.

She doesn't have a **debit card** or **credit card** of her own, so she asks her mom to pay online for her. Maricela also spent almost all her money at the mall, so she also has to borrow money from her mom for the present. Figure out how much Maricela has to borrow, and how she's going to pay for the present online on the following pages.

The tax on clothes in Maricela's state is 9.5%. This is a percent just like the sales percents you worked with before. However, this time it's a percent to add on to the cost of the tie, not a percent to subtract away.

1. How much does the tie cost with tax added on?

Now you have to add on the shipping and handling to get the full price of the tie bought online.

45

2. If the shipping and handling is $4.99, how much is the final price of the tie?

Maricela doesn't have that much money to give her mom yet.

3. Using your answer from section 5, how much money does Maricela need to get to pay her mom?

Maricela agrees to pay her mom back in **installments**. Every week, she will give her $5. That will give her enough time to save up money.

Use the chart below to figure out how many weeks it will take her to pay her mom back. Fill in the rest of the chart until you get to $0. You may not need to use all the rows.

Week	Original amount	Amount Left to Be Paid
1	$23.59	$18.59
2	$18.59	$13.59
3	$13.59	

4. How many weeks will it take Maricela to pay her mom back?

7
TAILORING

Once she tries on her new jeans at home again, Maricela realizes they don't quite fit. They fit her waist fine, but they are too long. She tries turning them up at the bottom, but they end up looking silly.

She doesn't want to return them. Besides, she already ripped the tag off. Her mom suggests she take them to the tailor around the corner. The tailor, her mom explains, will take her measurements, and then fix the jeans so they are the right length. Shortening pants is called hemming, in the tailoring world.

Maricela and her mom go to the tailor's and bring the jeans along. Her mom also brings a jacket that has gotten too tight. She doesn't want to get rid of it, and she thinks the tailor can **alter** it enough so it fits again.

The tailor, Mrs. Shah, brings Maricela and her mom to the back room. First she measures how long Maricela's legs are. Then she measures her mom's waist, where the jacket has gotten too tight. She tells them the jeans and the jacket will be ready the next day, and that they are easy fixes!

For the jeans, Mrs. Shah measures from the very top of the inseam (the seam that runs up the middle of the jeans, on the inside of the legs) to the bottom of the inseam. The bottom is where Maricela's pants will end. Maricela's inseam measures 28½ inches.

After they leave, Mrs. Shah measures the inseam of the jeans Maricela left. The inseam is 30¾ inches.

1. What is the difference between Maricela's actual leg length and the jeans?

This is how much Mrs. Shah will hem up the jeans. She wants to fold back the fabric so that the bottoms of the jeans are now at the right length.

Mrs. Shah will fold up the bottom of the jeans twice, so the original hem is completely hidden.

How many inches should she fold up each time? You will have to divide a fraction. Turn the total number of inches into an improper fraction. Then divide that number by two (because she is folding twice), and turn your answer back into a proper fraction.

2. What do you get?

Mrs. Shah folds up the hem and sews all around. Maricela's jeans are done!

Next, Mrs. Shah has to make Maricela's mom's jacket bigger by letting out the seams, the places where the sweater is sewed together.

Maricela's mom's waist was 41 inches around. When the jacket is buttoned, it is 41¼ inches around. The tailor has to make the jacket ¾ of an inch bigger, so that it won't feel so snug.

It's not actually as hard as it sounds. The tailor just has to rip out a seam and sew it back again, making sure she sews closer to the edge to give Maricela's mom more room. She will resew each side.

3. If she wants to let out the seams at total of ¾ of an inch, and she wants to let out that amount evenly on each of two sides, how much more room should each side have?

8
RESELLING OLD CLOTHES

While the tailor is fixing up the jeans, Maricela goes through her closet. She has a lot of clothes! She has barely even worn any of them. Those shirts in the back look great, she thinks—why hasn't she ever worn them? She takes them out and tries them on. They don't fit. She bought them a few years ago, and now they're too small for her.

Maricela starts a pile of clothes she wants to get rid of because they're too small, or she doesn't wear them anymore. She tosses in her old pair of jeans she just replaced, along with the shirts, and some other clothes.

She definitely doesn't want to throw all these clothes away, because someone can still wear them. Then she remembers her cousin telling her about reselling his old clothes to a **consignment** store. He brought his old clothes in, and the store took the ones that were in good enough shape. Then every time the store sold something he had brought in, it would give him a little money.

Maricela also remembers the used clothing store where she shops sometimes. You can bring old clothes there, and they pay you for the items that are in the best shape. The secondhand store is picky, but Maricela might have a couple things she could sell. Figure out how much she makes reselling her clothes on the next page.

Maricela first visits the second-hand store. She looks at the prices she could get for her clothes:

t-shirts: $3
button-up shirts: $4
sweaters: $5
pants: $5
skirts: $4
dresses: $6
jewelry: $3
jackets: $10

Maricela has brought in 3 pairs of pants, 4 t-shirts, 1 button-up shirt, 2 dresses, and 1 necklace.

1. What is the most she can make from the secondhand store?

The cashier looks over her things and accepts 1 pair of pants, 3 t-shirts, 1 dress, and the necklace.

2. How much did she make?

Next, Maricela heads over to the consignment store. It seems like it accepts clothes that aren't necessarily in perfect condition, so she hopes it will take everything she has left.

The consignment store gives people 20% of the price that the clothing sells for.

It ends up taking everything Maricela has. The woman there tells Maricela the prices she will use to sell each piece:

2 pairs of pants: $5.99 each
1 t-shirt: $2.99
1 button-up shirt: $4.50
1 dress: $8.75

3. How much will Maricela make if all these clothes are sold?

9
SEWING: CIRCUMFERENCE AND DIAMETER

Ever since she went to the tailor, Maricela has been thinking how cool it would be to sew her own clothes. She loves art, and has tried knitting and **crocheting**. She could save money and do something creative if she sewed her own clothes.

Maricela does some research and decides she wants to sew a simple circle skirt. She will need just one kind of fabric, and she will only have to make a few cuts. The basic idea is that she will cut out a circle. Then she will cut out a hole in the middle, as big as her waist. Then she will sew a waist on it, and hem the bottom.

The directions say she has to decide how long a skirt she wants. She chooses to make a skirt that is 18 inches long. Now she has to do some math to figure out how much fabric she needs to maker her skirt. Her goal is to find out the diameter—how wide the circle is—so she can figure how much cloth to buy.

For a skirt that is 18 inches long, she has to measure out 18 inches from the edge of the big circle to the outside edge of the little circle she will cut out in the middle.

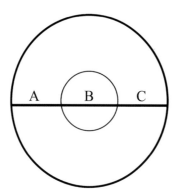

First she measures how wide her waist is. From the mall, she knows her waist is 38.5 inches around, which is the circumference of her waist. She can use that information to figure out how wide it is.

The formula for circumference is

$$C = \pi \times d$$

C is circumference, d is the diameter, and π is pi, or 3.14. You can rearrange the formula to give you diameter instead:

$$d = C/\pi$$

1. What is the diameter of Maricela's waist?

Now Maricela knows how big the hole in the middle should be (line B). Next she has to calculate how big the entire circle should be.

She wants the skirt to hang 18 inches down, so she has to make lines A and C at least 18 inches long. She also has to hem it, so she should add an extra 0.5 inch on either side.

Now she must add all the lengths together to find the total diameter of the circle, even with the hole in the middle.

2. What is the total diameter?

10 SEWING: RECTANGULAR DIMENSIONS

Maricela is very excited about her new hobby. Besides the skirt, she also wants to try designing and sewing a simple handbag. Her friend Jonathan at school made his own backpack. She doesn't think she's ready for that yet, but she thinks she could handle a small purse.

She looks at **patterns** online and sees that she could sew a bag made mostly out of rectangles sewn together. She wants to design her own bag to see if she really understands sewing. She takes out a piece of paper and starts drawing rectangles she will sew together into a bag, shown on the following page.

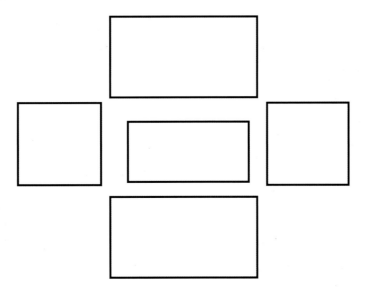

Maricela has her simple design, but she needs to figure out the **dimensions**. She has to know how big to cut the rectangles out of fabric so she can sew them together.

She starts with the center rectangle in her drawing, which will be the bottom of the bag. She just wants a small bag, so she makes the length (the longer end) 8 inches, and the width (the shorter end) 4 inches.

51

Then she realizes she has to add on a little bit for the seam, where it will get sewn to the other pieces. She adds ¼ inch all the way around, making the dimensions 8¼ by 4¼ inches.

The other pieces have to fit around the middle rectangle to make the sides. Maricela decides she wants the bag to be 5 inches tall, which will help her determine how big the other pieces are.

The smaller end pieces will have to be 5¼ inches long, to match how tall she wants the bag. They will also be 4¼ inches wide, to match the width of the middle rectangle.

1. How long should the larger end pieces be, and why?

2. How wide should the middle pieces be, and why?

She will also need a long, skinny rectangle for the handle, which will be 24 inches long and 2½ inches wide.

11
CHOOSING FABRIC

Maricela still needs to do a little more planning before she can actually start sewing her skirt and bag. She needs to figure out how much fabric she needs to buy. If she buys too much, then she wastes her money by buying something she doesn't need. If she buys too little, she won't be able to make her projects.

Figuring out the amount of fabric she needs means thinking about dimensions and also area, or how much room a shape takes up. Dimensions will help her out with the circle, while area is useful with her rectangular bag project. How will she figure it out?

Maricela knows the diameter of the big circle she will cut out for her skirt. You just calculated it in section 9.

Now think about the fabric. It comes on a roll that is 45 inches wide by however many yards long you need.

1. Is 45 inches wide big enough for the diameter of the circle? If so, how much extra fabric will there be?

Will a piece of fabric that is 45 inches by 1 yard be long enough? One way to figure it out is to see if 1 yard is longer than the diameter of the circle. Convert 1 yard to feet and then inches with this information:

$$1 \text{ yard} = 3 \text{ feet}, 1 \text{ foot} = 12 \text{ inches}$$

2. Is 1 yard long enough? If not, how many more inches will Maricela need?

Yards also come in quarters and halves. One-quarter yard is:

$$.25 \text{ yards} \times 3 \text{ feet} \times 12 \text{ inches} = 9 \text{ inches}.$$

3. Will a yard and a quarter fit the diameter of the circle?

Now Maricela needs to buy fabric for her bag. She knows the area of a rectangle is:

$$A = \text{length} \times \text{width}$$

She needs to add up all the areas of the rectangles she's going to cut out, and then get fabric with an area that is at least as large as that. Remember, ¼ inch is the same as saying .25 inches

4. What do all the areas of the rectangles add up to in square inches?

Now calculate the area in square inches of a piece of fabric that is 45 inches by ½ yard long:

5. 45 inches x (.5 yard x 3 feet x 12 inches) =

6. Is that enough fabric? If not, how much more does she need? If so, could she get ¼ yard and have enough?

12 FASHION BUSINESS: FIGURING OUT PRICES

Maricela discovers she really likes sewing clothes. Plus, she's pretty good at it. Her friends all really like her clothes, and soon they're asking if she can make them clothes too.

At first, Maricela really enjoys making clothes for her friends. She gets to practice, and she makes her friends happy. However, all the materials and supplies she has to buy really add up over time. She pays for everything herself, and she's going to be broke soon!

She's thinking about starting to sell her clothes to make some money. She might start selling at local craft fairs, or maybe set up a website and sell them online. Before she does that, she needs to figure out how much her clothes should be. She will have to add up all the costs of making them. To make a **profit**, she will need to price them higher than the cost of making them. Plus, she will need to figure out how much she thinks her time and skills are worth, and add that on to the price.

Maricela sets up a log to keep track of her sewing **expenses**. She creates one log for each sewing project.

Here's a log for her last project, a dress.

Item	Cost
Needles	$4.98
Thread, 5 spools	$7.45
Fabric	$26.23

1. What was her total cost for this project?

She takes another look at her costs. She didn't have to buy the needles or thread specifically for this project. She'll end up using both of those for many projects, not just this one. Her only cost specifically for the dress was the fabric.

2. How much do you think she needs at least to charge to make a profit on this dress?

Now she will calculate how much she thinks her time is worth, to get the whole price.

She spent 15 hours sewing the dress. Those 15 hours were spent sewing, not doing anything else. In fact, she could have been making money. Some of her friends have jobs at the mall, or at restaurants. They make about $9 an hour.

3. How much would Maricela have made in the time she sewed her dress if she were working at a store in the mall for $9 an hour?

Maricela adds how much money she could have been making to the cost of making the dress for a possible price. She takes a look to see if the price makes sense.

4. Do you think customers would pay that price for a dress? Is it too cheap or too expensive?

Maricela decides to make the dress cheaper, so that she can convince people to buy it. She is a new sewer after all. As her skills get better and better, she can charge higher prices.

She will charge a price that is 35% of the price she just calculated.

5. How much is the new price? Do you think it seems fair to customers and to Maricela?

13
SELLING ON COMMISSION

Maricela just heard about a new shop in town that sells people's artwork and crafts. They offer space for people to sell paintings, knit scarves, clothes, and more. The new store sells by commission. That is, they take some of the money from the sale, although most of it goes to the maker. Maricela never thought she would be able to sell her clothes in a store!

She goes and talks to the manager of the store, Tomás. Tomás tells her the shop takes a 15%

commission from the sale of clothes. In other words, for every piece of clothing she sells, the store will get 15% of it. Tomás tells her that the store will set the price of her clothes. They know what customers will spend, and can make prices to reflect that.

Maricela is a little nervous—what if the store sets the price really low? She doesn't need to worry, though, because Tomás tells her he will set the price at $75 a dress, $50 a shirt, $50 a skirt, and $40 a bag. Those are higher prices than she was selling her clothes at before! She is a really good fashion designer and sewer.

Finish filling out this chart to see how much money Maricela will make when her clothes sell:

Item	Price	Store Commission	Rest
Dress	$75	$11.25	$63.75
Shirt			
Skirt			
Bag			

1. Will Maricela make more money from selling her dress at the store than she was making before? If so, how much more? If not, how much less?

2. If Maricela spends $16.50 on fabric for a bag, what will her profit be?

3. How many shirts would she have to sew to make at least $150 in profit?

14
SELLING ONLINE

Selling clothes at the store is fun, but Maricela wants to reach more people. Only the customers who go in the store are seeing her clothes, but she thinks more people might be interested in buying them. When Maricela first started thinking about selling her clothes, she wanted to sell them online. Now is a good time to start!

First she considers setting up her own website, but she would have to work really hard to get people to come to her site in the first place. Another option is to sell her clothes on a **virtual marketplace**: a website someone else has set up for artists and crafters to sell what they make.

She signs up for a craft website and reads their rules. They will charge her $.20 for every thing she lists, and then take a 3.5% commission. She will have to come up with how much to charge in shipping and handling, to send the clothes to buyers. It seems like a lot of numbers! The next page will show how Maricela figures it all out.

First, Maricela wants to see how much she will make from online sales. She can set her own prices, so she decides to go with the prices the art shop set for her clothes.

How much money will she make, considering the costs she has to pay to the website? Take a look at a dress she will be selling for $75.

Costs:
$.20 for listing the dress online
3.5% x $75 = $2.63

1. How much will she make in total on the dress? How much of that will be profits if she spends $28.30 on the fabric?

Now calculate the shipping and handling using the following information:

- the dress weighs 15 ounces
- the box the dress will be packed in weighs 6 ounces
- the box costs $1.99
- the post office charges $.35 an ounce

How much should Maricela charge for shipping and handling for the dress?

total weight: 15 ounces + 6 ounces = 21 ounces
price based on weight: 21 ounces x $.35 = $7.35
price with box: $7.35 + $1.99 = 9.34

2. How much money will the customer have to pay including shipping and handling?

If someone in another country wants to buy the dress, the shipping and handling go up. The post office automatically charges $12, plus $.60 an ounce.

3. How much would an international customer pay in shipping and handling?

15
PUTTING IT ALL TOGETHER

Maricela has come a long way from her shopping trip at the mall. She has learned how clothes sizes work, how to keep track of the money she spends on clothes, how to design and sew clothes, and how to sell clothes to customers.

Someday, Maricela wants to go to school for fashion design and become a big name in fashion. She's well on her way by now, because she is learning so much math! See if you can remember what Maricela has learned along the way.

1. What size would a girl wear in U.S. junior sizes if she were a European size 32?

2. What size shoe would a boy wear if his feet (with socks) were 9¾ inches long?

3. While you are shopping for a sweatshirt, you find one that is on sale for 40% off. The original cost is $32.99. How much is it after the discount?

4. The sales tax on the sweatshirt is 9.5%. Can you afford it if you only have $25 to spend? Why or why not?

5. If a pair of pants is 29 inches, and the person who wants to wear them has an inseam of 26 inches, should the pants be hemmed?

   If so, by how much?

6. You give an old shirt to a consignment store. They price it at $6.50 and offer you 20%. How much will you get?

7. The area of the fabric you need to sew a shirt is 825 square inches. The craft store only has a half of a yard of the fabric you want to use. Will it be enough?

   If not, how much more fabric do you need?

8. You spend $8.80 on fabric for a sewing project. You sell it for $20.

   How much is your profit?

Answers

1.

1. (9 + 11 + 11 + 13 + 15)/5 = 11.8
2. 11
3. His average is 2, which is medium.
4. medium

2.

1. Size 14
2. 38
3. No, because they are only a size 26 in U.S. men's size.
4. A 52 in European size.

3.

1. 78
2. $78/8 - 3/8 = 75/8$
   $75/8 = 9 3/8$ inches
3. 9 ½
4. 8 ½

4.

1. $56 − ($24.93 + $6.70 + $6.70 + $6) = $11.67
2. $18.21
3. 2
4. one t-shirt. He will have $10.71 left if he buys one t-shirt.

5.

1. $5.50
2. $16.49. No, she doesn't have enough money.
3. $41.70
4. $13.90. No, she can't afford it.
5. Yes, she can buy it because it costs $10.20. She will have $2.47 left.

6.

1. $16.99 x .095 = $1.61, $16.99 + $1.61 = $18.60
2. $18.60 + $4.99 = $23.59
3. $23.59 − $2.47 = $21.12
4. 5 weeks

Week	Original amount	Amount Left to Be Paid
1	$23.59	$18.59
2	$18.59	$13.59
3	$13.59	$8.59
4	$8.59	$3.59
5	$3.59	<$0

7.

1. 2 ¼ inches
2. 1 1/8 inch each time (2 ¼ = 9/4, 9/4 / 2 = 9/8 = 1 1/8)
3. ¾ inch / 2 = 3/8 inch

8.

1. $46
2. $23
3. .20 x ((2 x $5.99) + $2.99 + $4.50 + $8.75) = $5.64

9.

1. d = 38.5 inches/3.14
   d = 12.26 inches
2. .5 inch + 18 inches + 12.26 inches + 18 inches + .5 inches = 49.26 inches

10.

1. 8 ¼ inches, to match the length of the middle piece.
2. 5 ¼ inches, to reach the same height as the rest of the pieces that make up the bag.

11.

1. No, it is not big enough.
2. No, she needs 13.26 more inches. (1 yard x 3 feet x 12 inches = 36 inches)
3. No
4. (2 x (5.25 x 4.25)) + (2 x (8.25 x 5.25) + (8.25 x 4.25) + (24 x 2.5) = 226.31 square inches
5. 810 square inches
6. Yes she has enough, yes she could get a ¼ yard, which is 405 square inches.

12.

1. $38.66
2. She will need to charge at least $26.24 to make a profit.
3. $135
4. $161.23. The dress seems expensive, and people might buy it.
5. $56.43. Yes, that seems fairer.

13.

1. She is making $7.32 more at the store
2. $17.50
3. 4 shirts

Item	Price	Store Commission	Rest
Dress	$75	$11.25	$63.75
Shirt	$50	$7.50	$42.50
Skirt	$50	$7.50	$42.50
Bag	$40	$6	$34

14.

1. $72.17, $43.87
2. $84.34
3. $24.60

15.

1. Junior's size 5.
2. 5
3. $19.80 (.4 x $32.99 = $13.19, $32.99 − $13.19 = $19.80)
4. Yes, you can afford it and have more than $3 left. ($19.80 x $0.095 = $1.88, $19.80 + $1.88 = $21.68)
5. Yes, by 3 inches.
6. $6.50 x .2 = $1.30
7. No, you need just .3 inches more.
   inches of fabric length:
   .5 yard x 3 feet x 12 inches = 18 inches
   length needed:
   825 square inches = length x 45 inches
   length = 18.3 inches
8. $11.20